Studies in Computational Intelligence

Volume 699

Series editor

Janusz Kacprzyk, Polish Academy of Sciences, Warsaw, Poland
e-mail: kacprzyk@ibspan.waw.pl

About this Series

The series "Studies in Computational Intelligence" (SCI) publishes new developments and advances in the various areas of computational intelligence—quickly and with a high quality. The intent is to cover the theory, applications, and design methods of computational intelligence, as embedded in the fields of engineering, computer science, physics and life sciences, as well as the methodologies behind them. The series contains monographs, lecture notes and edited volumes in computational intelligence spanning the areas of neural networks, connectionist systems, genetic algorithms, evolutionary computation, artificial intelligence, cellular automata, self-organizing systems, soft computing, fuzzy systems, and hybrid intelligent systems. Of particular value to both the contributors and the readership are the short publication timeframe and the worldwide distribution, which enable both wide and rapid dissemination of research output.

More information about this series at http://www.springer.com/series/7092

George A. Anastassiou

Intelligent Comparisons II: Operator Inequalities and Approximations

Springer

George A. Anastassiou
Department of Mathematical Sciences
University of Memphis
Memphis, TN
USA

ISSN 1860-949X ISSN 1860-9503 (electronic)
Studies in Computational Intelligence
ISBN 978-3-319-84660-6 ISBN 978-3-319-51475-8 (eBook)
DOI 10.1007/978-3-319-51475-8

Softcover reprint of the hardcover 1st edition 2017

Printed on acid-free paper

This Springer imprint is published by Springer Nature
The registered company is Springer International Publishing AG
The registered company address is: Gewerbestrasse 11, 6330 Cham, Switzerland

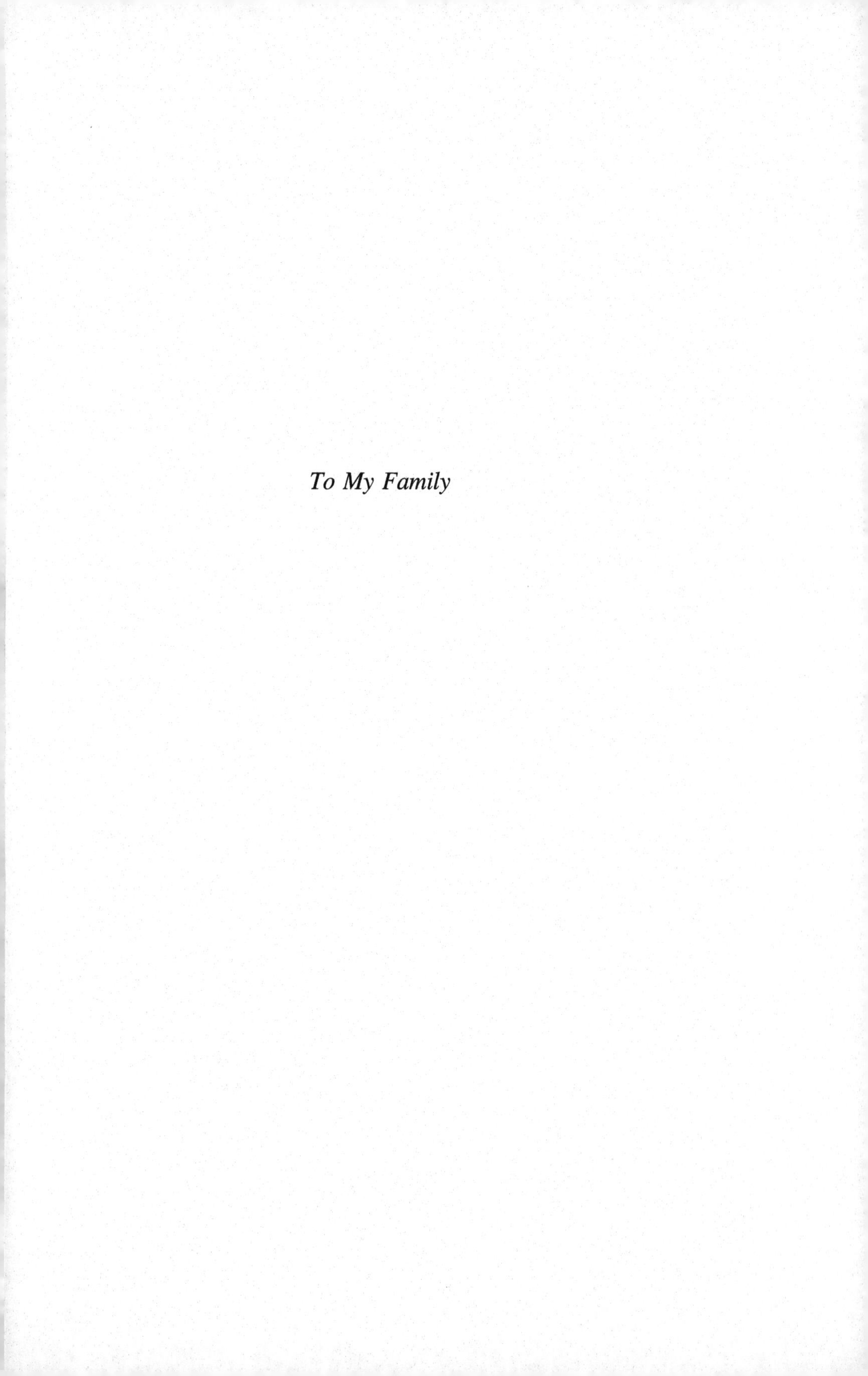

To My Family

Preface

This is a supplementary, complementary and companion brief monograph to the recently published monograph, by the same author, titled: "Intelligent Comparisons: Analytic Inequalities", Studies in Computational Intelligence 609, Springer Heidelberg/New York, 2016. It is the analog of the last one, regarding self-adjoint operator well-known inequalities and approximation theory of Korovkin type both in a Hilbert space environment. These are studied for the first time in the literature, and chapters are self-contained and can be read independently. This concise monograph is suitable to be used in related graduate classes and research projects.

The list of presented topics follows:

Self-adjoint operator Korovkin-type quantitative approximations.
Self-adjoint operator Korovkin type and polynomial direct approximations with rates.
Quantitative self-adjoint operator other direct approximations.
Fractional self-adjoint operator Poincaré- and Sobolev-type inequalities.
Self-adjoint operator Ostrowski-type inequalities.
Integer and fractional self-adjoint operator Opial-type inequalities.
Self-adjoint operator Chebyshev–Grüss-type inequalities.
Most general fractional self-adjoint operator representation formulae and operator Poincaré and Sobolev types and other basic inequalities.
Self-adjoint operator harmonic Chebyshev–Grüss inequalities.
Most general self-adjoint operator Chebyshev–Grüss inequalities.
A fractional means inequality.
An extensive list of references is given per chapter.

This book's results are expected to find applications in many areas of pure and applied mathematics. As such this monograph is suitable for researchers, graduate students, and seminars of the above disciplines, also to be in all science and engineering libraries.

The preparation of this book took place during 2016 in Memphis.

The author likes to thank Prof. Alina Lupas of University of Oradea, Romania, for checking and reading the manuscript.

Memphis, TN, USA
November 2016

George A. Anastassiou

Contents

Author Biography

George A. Anastassiou was born in Athens, Greece in 1952. He received his B.SC degree in Mathematics from Athens University, Greece in 1975. He received his Diploma in Operations Research from Southampton University, UK in 1976. He also received his MA in Mathematics from University of Rochester, USA in 1981. He was awarded his Ph. D in Mathematics from University of Rochester, USA in 1984. During 1984-86 he served as a visiting assistant professor at the University of Rhode Island, USA.

Since 1986 till now 2016, he is a faculty member at the University of Memphis, USA. He is currently a full Professor of Mathematics since 1994. His research area is "Computational Analysis" in the very broad sense. He has published over 426 research articles in international mathematical journals and 36 monographs, proceedings and textbooks in well-known publishing houses. Several awards have been awarded to George Anastassiou. In 2007 he received the Honorary Doctoral Degree from University of Oradea, Romania. He is associate editor in over 60 international mathematical journals and editor in-chief in 3 journals, most notably in the well-known "Journal of Computational Analysis and Applications".

Chapter 1
Self Adjoint Operator Korovkin Type Quantitative Approximation Theory

Here we present self adjoint operator Korovkin type theorems, via self adjoint operator Shisha-Mond type inequalities. This is a quantitative treatment to determine the degree of self adjoint operator uniform approximation with rates, of sequences of self adjoint operator positive linear operators. We give several applications involving the self adjoint operator Bernstein polynomials. It follows [2].

1.1 Background

Let A be a selfadjoint linear operator on a complex Hilbert space $(H; \langle \cdot, \cdot \rangle)$. The Gelfand map establishes a $*-$isometrically isomorphism Φ between the set $C(Sp(A))$ of all continuous functions defind on the spectrum of A, denoted $Sp(A)$, and the C^*-algebra $C^*(A)$ generated by A and the identity operator 1_H on H as follows (see e.g. [7, p. 3]):

For any $f, g \in C(Sp(A))$ and any $\alpha, \beta \in \mathbb{C}$ we have

(i) $\Phi(\alpha f + \beta g) = \alpha \Phi(f) + \beta \Phi(g)$;

(ii) $\Phi(fg) = \Phi(f)\Phi(g)$ (the operation composition is on the right) and $\Phi(\overline{f}) = (\Phi(f))^*$;

(iii) $\|\Phi(f)\| = \|f\| := \sup_{t \in Sp(A)} |f(t)|$;

(iv) $\Phi(f_0) = 1_H$ and $\Phi(f_1) = A$, where $f_0(t) = 1$ and $f_1(t) = t$, for $t \in Sp(A)$.

With this notation we define

$$f(A) := \Phi(f), \text{ for all } f \in C(Sp(A)),$$

and we call it the continuous functional calculus for a selfadjoint operator A.

G.A. Anastassiou, *Intelligent Comparisons II: Operator Inequalities and Approximations*, Studies in Computational Intelligence 699,
DOI 10.1007/978-3-319-51475-8_1

If A is a selfadjoint operator and f is a real valued continuous function on $Sp(A)$ then $f(t) \geq 0$ for any $t \in Sp(A)$ implies that $f(A) \geq 0$, i.e. $f(A)$ is a positive operator on H. Moreover, if both f and g are real valued continuous functions on $Sp(A)$ then the following important property holds:

(P) $f(t) \geq g(t)$ for any $t \in Sp(A)$, implies that $f(A) \geq g(A)$ in the operator order of $B(H)$.

Equivalently, we use (see [5], pp. 7–8):

Let U be a selfadjoint operator on the complex Hilbert space $(H, \langle \cdot, \cdot \rangle)$ with the spectrum $Sp(U)$ included in the interval $[m, M]$ for some real numbers $m < M$ and $\{E_\lambda\}_\lambda$ be its spectral family.

Then for any continuous function $f : [a, b] \to \mathbb{C}$, where [m, M] $\subset (a, b)$, it is well known that we have the following spectral representation in terms of the Riemann-Stieljes integral:

$$\langle f(U)x, y\rangle = \int_{m-0}^{M} f(\lambda)\, d(\langle E_\lambda x, y\rangle), \tag{1.1}$$

for any $x, y \in H$. The function $g_{x,y}(\lambda) := \langle E_\lambda x, y\rangle$ is of bounded variation on the interval $[m, M]$, and

$$g_{x,y}(m-0) = 0 \text{ and } g_{x,y}(M) = \langle x, y\rangle,$$

for any $x, y \in H$. Furthermore, it is known that $g_x(\lambda) := \langle E_\lambda x, x\rangle$ is increasing and right continuous on $[m, M]$.

In this chapter we will be using a lot the formula

$$\langle f(U)x, x\rangle = \int_{m-0}^{M} f(\lambda)\, d(\langle E_\lambda x, x\rangle), \ \forall\, x \in H. \tag{1.2}$$

As a symbol we can write

$$f(U) = \int_{m-0}^{M} f(\lambda)\, dE_\lambda. \tag{1.3}$$

Above, $m = \min\{\lambda | \lambda \in Sp(U)\} := \min Sp(U)$, $M = \max\{\lambda | \lambda \in Sp(U)\} := \max Sp(U)$. The projections $\{E_\lambda\}_{\lambda \in \mathbb{R}}$, are called the spectral family of A, with the properties:

(a) $E_\lambda \leq E_{\lambda'}$ for $\lambda \leq \lambda'$;
(b) $E_{m-0} = 0_H$ (zero operator), $E_M = 1_H$ (identity operator) and $E_{\lambda+0} = E_\lambda$ for all $\lambda \in \mathbb{R}$.

Furthermore

$$E_\lambda := \varphi_\lambda(U), \ \forall\, \lambda \in \mathbb{R}, \tag{1.4}$$

is a projection which reduces U, with

$$\varphi_\lambda(s) := \begin{cases} 1, & \text{for } -\infty < s \leq \lambda, \\ 0, & \text{for } \lambda < s < +\infty. \end{cases}$$

The spectral family $\{E_\lambda\}_{\lambda \in \mathbb{R}}$ determines uniquely the self-adjoint operator U and vice versa.

For more on the topic see [6], pp. 256–266, and for more detalis see there pp. 157–266.

Some more basics are given (we follow [5], pp. 1–5):

Let $(H; \langle \cdot, \cdot \rangle)$ be a Hilbert space over $\mathbb{C}$. A bounded linear operator A defined on H is selfjoint, i.e., $A = A^*$, iff $\langle Ax, x \rangle \in \mathbb{R}$, $\forall\, x \in H$, and if A is selfadjoint, then

$$\|A\| = \sup_{x \in H: \|x\|=1} |\langle Ax, x \rangle| . \tag{1.5}$$

Let A, B be selfadjoint operators on H. Then $A \leq B$ iff $\langle Ax, x \rangle \leq \langle Bx, x \rangle, \forall\, x \in H$.

In particular, A is called positive if $A \geq 0$.

Denote by

$$\mathcal{P} := \left\{ \varphi(s) := \sum_{k=0}^{n} \alpha_k s^k | n \geq 0, \alpha_k \in \mathbb{C}, 0 \leq k \leq n \right\}. \tag{1.6}$$

If $A \in \mathcal{B}(H)$ (the Banach algebra of all bounded linear operators defined on H, i.e. from H into itself) is selfadjoint, and $\varphi(s) \in \mathcal{P}$ has real coefficients, then $\varphi(A)$ is selfadjoint, and

$$\|\varphi(A)\| = \max\{|\varphi(\lambda)|, \lambda \in Sp(A)\}. \tag{1.7}$$

If φ is any function defined on $\mathbb{R}$ we define

$$\|\varphi\|_A := \sup\{|\varphi(\lambda)|, \lambda \in Sp(A)\}. \tag{1.8}$$

If A is selfadjoint operator on Hilbert space H and φ is continuous and given that $\varphi(A)$ is selfadjoint, then $\|\varphi(A)\| = \|\varphi\|_A$. And if φ is a continuous real valued function so it is $|\varphi|$, then $\varphi(A)$ and $|\varphi|(A) = |\varphi(A)|$ are selfadjoint operators (by [5], p. 4, Theorem 7).

Hence it holds

$$\begin{aligned} \| |\varphi(A)| \| &= \| |\varphi| \|_A = \sup\{||\varphi(\lambda)||, \lambda \in Sp(A)\} \\ &= \sup\{|\varphi(\lambda)|, \lambda \in Sp(A)\} = \|\varphi\|_A = \|\varphi(A)\|, \end{aligned}$$

that is

$$\| |\varphi(A)| \| = \|\varphi(A)\|. \tag{1.9}$$

For a selfadjoint operator $A \in \mathcal{B}(H)$ which is positive, there exists a unique positive selfadjoint operator $B := \sqrt{A} \in \mathcal{B}(H)$ such that $B^2 = A$, that is $\left(\sqrt{A}\right)^2 = A$. We call B the square root of A.

Let $A \in \mathcal{B}(H)$, then A^*A is selfadjoint and positive. Define the "operator absolute value" $|A| := \sqrt{A^*A}$. If $A = A^*$, then $|A| = \sqrt{A^2}$.

For a continuous real valued function φ we observe the following:

$$|\varphi(A)| \text{ (the functional absolute value)} = \int_{m-0}^{M} |\varphi(\lambda)| \, dE_\lambda =$$

$$\int_{m-0}^{M} \sqrt{(\varphi(\lambda))^2} dE_\lambda = \sqrt{(\varphi(A))^2} = |\varphi(A)| \text{ (operator absolute value)},$$

where A is a selfadjoint operator.

That is we have

$$|\varphi(A)| \text{ (functional absolute value)} = |\varphi(A)| \text{ (operator absolute value)}. \quad (1.10)$$

The next comes from [4], p. 3:

We say that a sequence $\{A_n\}_{n=1}^{\infty} \subset \mathcal{B}(H)$ converges uniformly to A (convergence in norm), iff

$$\lim_{n\to\infty} \|A_n - A\| = 0, \quad (1.11)$$

and we denote it as $\lim_{n\to\infty} A_n = A$.

We will be using Hölder's-McCarthy, 1967 ([8]), inequality: Let A be a selfadjoint positive operator on a Hilbert space H. Then

$$\left\langle A^r x, x \right\rangle \leq \langle Ax, x \rangle^r, \quad (1.12)$$

for all $0 < r < 1$ and $x \in H : \|x\| = 1$.

Let $A, B \in \mathcal{B}(H)$, then

$$\|AB\| \leq \|A\| \, \|B\|, \quad (1.13)$$

by Banach algebra property.

1.2 Auxiliary Results

All functions here are real valued.

Let $L : C([a, b]) \to C([a, b])$, $a < b$, be a linear operator. If $f, g \in C([a, b])$ such that $f \geq g$ implies $L(f) \geq L(g)$, we call L a positive linear operator. It is well-known that a positive linear operator is a bounded linear operator.

We need

Lemma 1.1 *Let $L : C([a, b]) \to C([a, b])$ be a positive linear operator,* $0 < \alpha \leq 1$. *Then the function*

$$g(x) := (L(|\cdot - x|^{\alpha}))(x) \tag{1.14}$$

is continuous in $x \in [a, b]$.

Proof Let $x_n \to x$, $x_n, x \in [a, b]$. We notice that

$$(L(|\cdot - x_n|^{\alpha}))(x_n) - (L(|\cdot - x|^{\alpha}))(x) =$$

$$(L(|\cdot - x_n|^{\alpha}))(x_n) - (L(|\cdot - x|^{\alpha}))(x_n) + (L(|\cdot - x|^{\alpha}))(x_n) - (L(|\cdot - x|^{\alpha}))(x) =$$

$$(L(|\cdot - x_n|^{\alpha} - |\cdot - x|^{\alpha}))(x_n) + \left[(L(|\cdot - x|^{\alpha}))(x_n) - (L(|\cdot - x|^{\alpha}))(x)\right]. \tag{1.15}$$

Therefore it holds

$$|(L(|\cdot - x_n|^{\alpha}))(x_n) - (L(|\cdot - x|^{\alpha}))(x)| \leq$$

$$\|L(|\cdot - x_n|^{\alpha} - |\cdot - x|^{\alpha})\|_{\infty} + |(L(|\cdot - x|^{\alpha}))(x_n) - (L(|\cdot - x|^{\alpha}))(x)| \leq \tag{1.16}$$

$$\|L\| \, \||\cdot - x_n|^{\alpha} - |\cdot - x|^{\alpha}\|_{\infty} + |(L(|\cdot - x|^{\alpha}))(x_n) - (L(|\cdot - x|^{\alpha}))(x)| =: (\xi_1). \tag{1.17}$$

Notice that

$$|t - x_n| = |t - x + x - x_n| \leq |t - x| + |x - x_n|,$$

hence

$$|t - x_n|^{\alpha} \leq (|t - x| + |x - x_n|)^{\alpha} \leq |t - x|^{\alpha} + |x - x_n|^{\alpha}.$$

That is

$$|t - x_n|^{\alpha} - |t - x|^{\alpha} \leq |x - x_n|^{\alpha}. \tag{1.18}$$

Similarly

$$|t - x| = |t - x_n + x_n - x| \leq |t - x_n| + |x_n - x|,$$

hence

$$|t - x|^{\alpha} \leq |t - x_n|^{\alpha} + |x_n - x|^{\alpha},$$

and

$$|t - x|^{\alpha} - |t - x_n|^{\alpha} \leq |x_n - x|^{\alpha}. \tag{1.19}$$

Consequently, it holds

$$||t - x_n|^{\alpha} - |t - x|^{\alpha}| \leq |x_n - x|^{\alpha}, \tag{1.20}$$

and

$$\| |\cdot - x_n|^{\alpha} - |\cdot - x|^{\alpha} \|_{\infty} \leq |x_n - x|^{\alpha} . \tag{1.21}$$

Therefore we get

$$(\xi_1) \leq \|L\| \, |x_n - x|^{\alpha} + |(L(|\cdot - x|^{\alpha}))(x_n) - (L(|\cdot - x|^{\alpha}))(x)| \to 0, \text{ as } x_n \to x, \tag{1.22}$$

and by continuity of $(L(|\cdot - x|^{\alpha}))$, as $n \to \infty$, proving the claim. ■

We make

Remark 1.2 Let L be a positive linear operator from $C([a, b])$ into itself. Then

$$(t - x)^n = \sum_{k=0}^{n} (-1)^{n-k} \binom{n}{k} t^k x^{n-k}, \; t, x \in [a, b] . \tag{1.23}$$

Hence we get

$$\left(L\left((t - x)^n\right)\right) = \sum_{k=0}^{n} (-1)^{n-k} \binom{n}{k} x^{n-k} L\left(t^k\right), \tag{1.24}$$

and

$$\left(L\left((t - x)^n\right)\right)(x) = \sum_{k=0}^{n} (-1)^{n-k} \binom{n}{k} x^{n-k} \left(L\left(t^k\right)\right)(x), \; \forall \, x \in [a, b]. \tag{1.25}$$

Clearly we have that $\left(L\left((\cdot - x)^n\right)\right)(x)$ is continuous in x, $\forall \, n \in \mathbb{N}$. So that $\left|\left(L\left((\cdot - x)^n\right)\right)(x)\right|$ is continuous in $x \in [a, b]$.

It follows

Lemma 1.3 *Let L be a positive linear operator from $C([a, b])$ into itself. The function $(L(|\cdot - x|^m))(x)$ is continuous in $x \in [a, b]$, for any $m \in \mathbb{N}$.*

Proof Let $x_n \to x$, $x_n, x \in [a, b]$, as $n \to \infty$.
We observe that

$$\left\| L\left(|\cdot - x_n|^m - |\cdot - x|^m\right) \right\|_{\infty} \leq \|L\| \, \left\| |\cdot - x_n|^m - |\cdot - x|^m \right\|_{\infty} . \tag{1.26}$$

We notice that $(t, x_n, x \in [a, b])$

$$\left| |t - x_n|^m - |t - x|^m \right| = \left| |t - x_n| - |t - x| \right| \left\{ |t - x_n|^{m-1} + |t - x_n|^{m-2} |t - x| \right.$$

$$\left. + |t - x_n|^{m-3} |t - x|^2 + ... + |t - x_n| \, |t - x|^{m-2} + |t - x|^{m-1} \right\} \leq \tag{1.27}$$

$$||t - x_n| - |t - x|| \, m \, (b - a)^{m-1} \leq |x_n - x| \, m \, (b - a)^{m-1}.$$

Hence it holds

$$\left\| |\cdot - x_n|^m - |\cdot - x|^m \right\|_\infty \leq |x_n - x| \, m \, (b - a)^{m-1}. \tag{1.28}$$

Similarly, as in the proof of Lemma 1.1 (instead of α we set m), we obtain

$$\left| \left(L \left(|\cdot - x_n|^m \right) \right) (x_n) - \left(L \left(|\cdot - x|^m \right) \right) (x) \right| \leq$$

$$\|L\| \left\| |\cdot - x_n|^m - |\cdot - x|^m \right\|_\infty + \left| \left(L \left(|\cdot - x|^m \right) \right) (x_n) - \left(L \left(|\cdot - x|^m \right) \right) (x) \right| \leq \tag{1.29}$$

$$\|L\| \, |x_n - x| \, m \, (b - a)^{m-1} + \left| \left(L \left(|\cdot - x|^m \right) \right) (x_n) - \left(L \left(|\cdot - x|^m \right) \right) (x) \right| \to 0, \tag{1.30}$$

proving the claim. ■

We also need

Lemma 1.4 *Let L be a positive linear operator from $C([a, b])$ into itself. The function $\left(L \left(|\cdot - x|^{n+\alpha} \right) \right) (x)$ is continuous in $x \in [a, b]$, $n \in \mathbb{N}$, $0 < \alpha \leq 1$.*

Proof Let $0 \leq A, B \leq b - a$, and $\gamma (z) := z^r$, $r > 1$, with $\gamma : [0, b - a] \to \mathbb{R}$, i.e. $\gamma (A) = A^r$, $\gamma (B) = B^r$. Then $\gamma' (z) = r z^{r-1}$, and $\|\gamma'\|_\infty = r (b - a)^{r-1}$.

Hence it holds

$$\left| A^r - B^r \right| \leq r (b - a)^{r-1} |A - B|. \tag{1.31}$$

Let $t, x_m, x \in [a, b]$, with $x_m \to x$, as $m \to \infty$.

Therefore (for $r = n + \alpha > 1$, $A = |t - x_m|$, $B = |t - x|$) we get that

$$\left| |t - x_m|^{n+\alpha} - |t - x|^{n+\alpha} \right| \leq (n + \alpha) (b - a)^{n+\alpha-1} \, ||t - x_m| - |t - x|| \tag{1.32}$$

$$\leq (n + \alpha) (b - a)^{n+\alpha-1} |x_m - x|.$$

So that it holds

$$\left\| |t - x_m|^{n+\alpha} - |t - x|^{n+\alpha} \right\|_\infty \leq (n + \alpha) (b - a)^{n+\alpha-1} |x_m - x| \to 0. \tag{1.33}$$

We have that

$$\left| \left(L \left(|\cdot - x_m|^{n+\alpha} \right) \right) (x_m) - \left(L \left(|\cdot - x|^{n+\alpha} \right) \right) (x) \right| \leq$$

$$\|L\| \left\| |\cdot - x_m|^{n+\alpha} - |\cdot - x|^{n+\alpha} \right\|_\infty +$$

$$\left| \left(L \left(|\cdot - x|^{n+\alpha} \right) \right) (x_m) - \left(L \left(|\cdot - x|^{n+\alpha} \right) \right) (x) \right| \leq$$

$$\|L\| \, |x_m - x| \, (n + \alpha) (b - a)^{n+\alpha-1} + \tag{1.34}$$

$$\left|\left(L\left(|\cdot-x|^{n+\alpha}\right)\right)(x_m)-\left(L\left(|\cdot-x|^{n+\alpha}\right)\right)(x)\right|\to 0,$$

proving the claim. ∎

We make

Remark 1.5 Let L be a positive linear operator from $C([a,b])$ into itself, $a<b$. By Riesz representation theorem, for each $s\in[a,b]$, there exists a positive finite measure μ_s on $[a,b]$ such that

$$(L(f))(s)=\int_{[a,b]} f(t)\,d\mu_s(t),\ \forall\ f\in C([a,b]). \tag{1.35}$$

Therefore ($k=1,...,n;\ 0<\alpha\leq 1$)

$$\left|\left(L(\cdot-s)^k\right)(s)\right|=\left|\int_{[a,b]}(\lambda-s)^k\,d\mu_s(\lambda)\right|$$

$$\leq\int_{[a,b]}|\lambda-s|^k\,d\mu_s(\lambda)\ \ \text{(by Hölder's inequality)}$$

$$\leq\left(\int_{[a,b]}1d\mu_s(\lambda)\right)^{\frac{((n+\alpha)-k)}{n+\alpha}}\left(\int_{[a,b]}|\lambda-s|^{(n+\alpha)}\,d\mu_s(\lambda)\right)^{\frac{k}{n+\alpha}} \tag{1.36}$$

$$=\left((L(1))(s)\right)^{\left(\frac{n+\alpha-k}{n+\alpha}\right)}\left(\left(L\left(|\cdot-s|^{n+\alpha}\right)\right)(s)\right)^{\frac{k}{n+\alpha}}.$$

We have proved that ($k=1,...,n;\ 0<\alpha\leq 1$)

$$\left|\left(L(\cdot-s)^k\right)(s)\right|\leq\left((L(1))(s)\right)^{\left(\frac{n+\alpha-k}{n+\alpha}\right)}\left(\left(L\left(|\cdot-s|^{n+\alpha}\right)\right)(s)\right)^{\frac{k}{n+\alpha}},\ \forall\ s\in[a,b]. \tag{1.37}$$

We mention

Theorem 1.6 (Shisha and Mond ([9]), 1968) *Let $[a,b]\subset\mathbb{R}$ be a compact interval. Let $\{L_n\}_{n\in\mathbb{N}}$ be a sequence of positive linear operators acting from $C([a,b])$ into itself. For $n=1,2,...,$ suppose $L_n(1)$ is bounded. Let $f\in C([a,b])$. Then for $n=1,2,...,$ we have*

$$\|L_nf-f\|_\infty\leq\|f\|_\infty\|L_n1-1\|_\infty+\|L_n(1)+1\|_\infty\,\omega_1(f,\mu_n), \tag{1.38}$$

where

$$\mu_n:=\left\|\left(L_n\left((t-x)^2\right)\right)(x)\right\|_\infty^{\frac{1}{2}}, \tag{1.39}$$

with

$$\omega_1(f,\delta) := \sup_{\substack{x,y\in[a,b]:\\|x-y|\le\delta}} |f(x) - f(y)|, \ \delta > 0, \tag{1.40}$$

and $\|\cdot\|_\infty$ *stands for the sup-norm over* $[a,b]$. *In particular, if* $L_n(1) = 1$, *then (1.38) becomes*

$$\|L_n(f) - f\|_\infty \le 2\omega_1(f,\mu_n). \tag{1.41}$$

Note: (i) In forming μ_n^2, x is kept fixed, however t forms the functions t, t^2 on which L_n acts.

(ii) One can easily find, for $n = 1, 2, \dots$,

$$\mu_n^2 \le \left\|\left(L_n\left(t^2\right)\right)(x) - x^2\right\|_\infty + 2c\,\|(L_n(t))(x) - x\|_\infty + c^2\,\|(L_n(1))(x) - 1\|_\infty, \tag{1.42}$$

where $c := \max(|a|, |b|)$.

So, if the Korovkin's assumptions are fulfilled, i.e. if $L_n\left(id^2\right) \xrightarrow{u} id^2$, $L_n(id) \xrightarrow{u} id$ and $L_n(1) \xrightarrow{u} 1$, as $n \to \infty$, where id is the identity map and u is the uniform convergence, then $\mu_n \to 0$, and then $\omega_1(f,\mu_n) \to 0$, as $n \to +\infty$, and we obtain from (1.38) that $\|L_n f - f\|_\infty \to 0$, i.e. $L_n f \xrightarrow{u} f$, as $n \to \infty$, $\forall\, f \in C([a,b])$.

Clearly the assumption $\|L_n(1) - 1\|_\infty \to 0$, as $n \to \infty$, implies $\|L_n(1)\|_\infty \le \rho$, $\forall\, n \in \mathbb{N}$, for some $\rho > 0$.

Indeed we can write $L_n(1) = L_n(1) - 1 + 1$, hence

$$\|L_n(1)\|_\infty \le \|L_n(1) - 1\|_\infty + 1 \le \rho,$$

proving the boundedness of $L_n(1)$.

1.3 Main Results

Here we derive self adjoint operator-Korovkin type theorems via operator-Shisha-Mond type inequalities. This is a quantitative approach, studying the degree of operator-uniform approximation with rates of sequences of operator-positive linear operators in the operator order of $\mathcal{B}(H)$.

In all of our results here we give direct self contained proofs by the use of spectral representation theorem.

We are inspired by [1].

Our setting here follows:

Let A be a selfadjoint operator on the Hilbert space H with the spectrum $Sp(A) \subseteq [m,M]$ for some real numbers $m < M$, $\{E_\lambda\}_\lambda$ be its spectral family, $I = [a,b]$, $a < b$, a, b real numbers, with $[m,M] \subset \overset{\circ}{I} = (a,b)$ (the interior of I). Let $f \in C(I)$, where $C(I)$ denotes all the continuous functions from I into $\mathbb{R}$. Let $n \in \mathbb{N}$ and $\{L_n\}_{n\in\mathbb{N}}$ be a sequence of positive linear operators from $C(I)$ into itself.

We give

Theorem 1.7 *It holds*

$$\|(L_n(f))(A) - f(A)\| \leq \|L_n f - f\|_{\infty,[a,b]}, \ \forall n \in \mathbb{N}. \tag{1.43}$$

If $L_n 1 \overset{u}{\to} 1$, $L_n(id) \overset{u}{\to} id$, $L_n\left(id^2\right) \overset{u}{\to} id^2$, *then* $\|L_n f - f\|_{\infty,[a,b]} \underset{[a,b]}{\to} 0$, *(see Theorem 1.6 and Note).*

By (1.43), then we get $\|(L_n(f))(A) - f(A)\| \to 0$, *as* $n \to \infty$, *i.e.*

$$\lim_{n\to\infty} (L_n(f))(A) = f(A), \text{ uniformly}, \forall f \in C(I).$$

Proof Here we use the spectral representation theorem.

For any $x \in H : \|x\| = 1$, we have that $\int_{m-0}^{M} d\langle E_\lambda x, x\rangle = 1$, see (1.1) and (1.2).

We observe that

$$\|(L_n(f))(A) - f(A)\| = \sup_{\substack{x\in H \\ \|x\|=1}} |\langle((L_n(f))(A) - f(A))x, x\rangle| = \tag{1.44}$$

$$\sup_{\|x\|=1} \left| \int_{m-0}^{M} ((L_n(f))(\lambda) - f(\lambda))\, d\langle E_\lambda x, x\rangle \right| \leq$$

$$\sup_{\|x\|=1} \int_{m-0}^{M} |(L_n(f))(\lambda) - f(\lambda)|\, d\langle E_\lambda x, x\rangle \leq$$

$$\|L_n(f) - f\|_{\infty,[a,b]} \left(\sup_{\|x\|=1} \int_{m-0}^{M} d\langle E_\lambda x, x\rangle \right) =$$

$$\|L_n(f) - f\|_{\infty,[a,b]} \cdot 1 = \|L_n(f) - f\|_{\infty,[a,b]}, \tag{1.45}$$

proving the claim. ∎

Next we give special Korovkin type quantitative convergence results for a self adjoint operator A.

We present

Theorem 1.8 *Let* $f : [a,b] \to \mathbb{R}$. *Assume that*

$$|f(t) - f(s)| \leq K|t - s|^\alpha, \forall t, s \in [a,b], \tag{1.46}$$

where $0 < \alpha \leq 1$, $K > 0$.

Assume that

$$\|L_n(1)\|_{\infty,[a,b]} \leq \mu,\ \mu > 0,\ \forall\, n \in \mathbb{N}, \tag{1.47}$$

and set

$$\rho := \mu^{\frac{2-\alpha}{2}}. \tag{1.48}$$

Set also $c := \max(|a|, |b|)$.

Then it holds

$$\|(L_n(f))(A) - f(A)\| \leq \|f(A)\| \, \|(L_n(1))(A) - 1_H\|$$

$$+ K\rho \left[\left\| \left(L_n\left(id^2\right)\right)(A) - A^2 \right\| + 2c\,\|(L_n(id))(A) - A\| + c^2\,\|(L_n(1))(A) - 1_H\| \right]^{\frac{\alpha}{2}}, \tag{1.49}$$

$\forall\, n \in \mathbb{N}$.

If we assume that $\left(L_n\left(id^2\right)\right)(A) \to A^2$, $(L_n(id))(A) \to A$, $(L_n(1))(A) \to 1_H$, *uniformly, as* $n \to \infty$, *we get* $(L_n(f))(A) \to f(A)$ *uniformly, as* $n \to \infty$, $\forall\, f \in C([a,b])$ *fulfilling (1.46).*

Proof Here we consider the sequence of positive linear operators $\{L_n\}_{n\in\mathbb{N}}$ from $C([a,b])$ into itself. By Riesz representation theorem we have that

$$(L_n(f))(s) = \int_{[a,b]} f(t)\,\mu_{ns}(dt), \tag{1.50}$$

$\forall\, f \in C([a,b])$; where μ_{ns} is a non-negative finite measure, $\forall\, s \in [a,b]$; $\forall\, n \in \mathbb{N}$.

We can write the following

$$(L_n(f))(s) - f(s) = (L_n(f))(s) - f(s) + f(s)(L_n(1))(s) - f(s)(L_n(1))(s) = \tag{1.51}$$

$$\int_{[a,b]} (f(t) - f(s))\,\mu_{ns}(dt) + f(s)((L_n(1))(s) - 1).$$

By the assumption (1.46) we obtain

$$|(L_n(f))(s) - f(s)| \leq \int_{[a,b]} |f(t) - f(s)|\,\mu_{ns}(dt) + |f(s)|\,|(L_n(1))(s) - 1| \leq \tag{1.52}$$

$$K \int_{[a,b]} |t - s|^{\alpha}\,\mu_{ns}(dt) + |f(s)|\,|(L_n(1))(s) - 1| =$$

$$K\left(L_n\left(|\cdot - s|^{\alpha}\right)(s)\right) + |f(s)|\,|(L_n(1))(s) - 1|.$$

That is, we get

$$|(L_n(f))(s) - f(s)| \leq |f(s)| |(L_n(1))(s) - 1| + K(L_n(|\cdot - s|^\alpha)(s)). \quad (1.53)$$

Notice that (by Hölder's inequality)

$$(L_n(|\cdot - s|^\alpha)(s)) = \int_{[a,b]} |t - s|^\alpha \mu_{ns}(dt)$$

$$\leq ((L_n(1))(s))^{\frac{2-\alpha}{2}} \left(\int_{[a,b]} (t-s)^2 \mu_{ns}(dt) \right)^{\frac{\alpha}{2}}. \quad (1.54)$$

Hence it holds

$$|(L_n(f))(s) - f(s)| \leq |f(s)| |(L_n(1))(s) - 1|$$

$$+ K((L_n(1))(s))^{\frac{2-\alpha}{2}} \left((L_n((\cdot - s)^2))(s) \right)^{\frac{\alpha}{2}}. \quad (1.55)$$

By the assumption (1.47) and (1.48) we get

$$|(L_n(f))(s) - f(s)| \leq |f(s)| |(L_n(1))(s) - 1| + K\rho \left((L_n((\cdot - s)^2))(s) \right)^{\frac{\alpha}{2}}, \quad (1.56)$$

$\forall s \in [a, b]$.

We see that

$$\left(L_n((t-s))^2\right)(s) = \left(L_n\left(t^2 - 2ts + s^2\right)\right)(s) =$$

$$\left(L_n\left(t^2\right)\right)(s) - 2s(L_n(t))(s) + s^2(L_n(1))(s) =$$

$$\left(\left(L_n\left(t^2\right)\right)(s) - s^2\right) - 2s((L_n(t))(s) - s) + s^2((L_n(1))(s) - 1). \quad (1.57)$$

Calling $c := \max(|a|, |b|)$, we obtain

$$\left(L_n\left((t-s)^2\right)\right)(s) \leq \left|\left(L_n\left(t^2\right)\right)(s) - s^2\right| + \quad (1.58)$$

$$2c|(L_n(t))(s) - s| + c^2|(L_n(1))(s) - 1|.$$

Therefore

$$|(L_n(f))(s) - f(s)| \leq |f(s)| |(L_n(1))(s) - 1| + \quad (1.59)$$

$$K\rho\left[\left|\left(L_n\left(t^2\right)\right)(s) - s^2\right| + 2c|(L_n(t))(s) - s| + c^2|(L_n(1))(s) - 1|\right]^{\frac{\alpha}{2}},$$

$\forall s \in [a, b]; \forall n \in \mathbb{N}$.

Here we take $x \in H : \|x\| = 1$. We find that

$$|\langle((L_n(f))(A) - f(A))x, x\rangle| = \left|\int_{m-0}^{M} ((L_n(f))(s) - f(s))\, d\langle E_s x, x\rangle\right| \leq \tag{1.60}$$

$$\int_{m-0}^{M} |(L_n(f))(s) - f(s)|\, d\langle E_s x, x\rangle \leq \int_{m-0}^{M} |f(s)|\, |(L_n(1))(s) - 1|\, d\langle E_s x, x\rangle + \tag{1.61}$$

$$K\rho \int_{m-0}^{M} \left[\left|\left(L_n\left(t^2\right)\right)(s) - s^2\right| + 2c\,|(L_n(t))(s) - s| + c^2\,|(L_n(1))(s) - 1|\right]^{\frac{\alpha}{2}}.$$

$$d\langle E_s x, x\rangle \overset{(1.10)}{=} \langle(|f(A)|\,|(L_n(1))(A) - 1_H|)x, x\rangle +$$

$$K\rho\left\langle\left[\left|\left(L_n\left(t^2\right)\right)(A) - A^2\right| + 2c\,|(L_n(t))(A) - A| + c^2\,|(L_n(1))(A) - 1_H|\right]^{\frac{\alpha}{2}} x, x\right\rangle \tag{1.62}$$

(by Hölder-McCarthy inequality (1.12) and (1.5), (1.9) and (1.13))

$$\leq \|f(A)\|\,\|(L_n(1))(A) - 1_H\| + K\rho\cdot$$

$$\left(\left\langle\left[\left|\left(L_n\left(id^2\right)\right)(A) - A^2\right| + 2c\,|(L_n(id))(A) - A| + c^2\,|(L_n(1))(A) - 1_H|\right]x, x\right\rangle\right)^{\frac{\alpha}{2}} \tag{1.63}$$

$$= \|f(A)\|\,\|(L_n(1))(A) - 1_H\| + K\rho\left(\left\langle\left|\left(L_n\left(id^2\right)\right)(A) - A^2\right| x, x\right\rangle +\right.$$

$$\left.2c\,\langle|(L_n(id))(A) - A|\,x, x\rangle + c^2\,\langle|(L_n(1))(A) - 1_H|\,x, x\rangle\right)^{\frac{\alpha}{2}} \leq \tag{1.64}$$

$$\|f(A)\|\,\|(L_n(1))(A) - 1_H\| +$$

$$K\rho\left(\left\|\left(L_n\left(id^2\right)\right)(A) - A^2\right\| + 2c\,\|(L_n(id))(A) - A\| + c^2\,\|(L_n(1))(A) - 1_H\|\right)^{\frac{\alpha}{2}}, \tag{1.65}$$

proving (1.49). ■

It follows a related result.

Theorem 1.9 *Let $f : [a, b] \to \mathbb{R}$. Assume that*

$$|f(t) - f(s)| \leq K\,|t - s|^{\alpha}, \ \forall\, t, s \in [a, b], \tag{1.66}$$

where $0 < \alpha \leq 1$, $K > 0$.

Then

$$\|(L_n(f))(A) - f(A)\| \leq$$

$$\|f(A)\| \, \|(L_n(1))(A) - 1_H\| + K \, \|(L_n(|\cdot - A|^{\alpha}))(A)\|, \; \forall \, n \in \mathbb{N}. \quad (1.67)$$

Clearly, if $(L_n(1))(A) \to 1_H$ *and* $(L_n(|\cdot - A|^{\alpha}))(A) \to 0_H$, *uniformly as* $n \to \infty$, *then by (1.67) we get that* $(L_n(f))(A) \to f(A)$, *uniformly, as* $n \to \infty$.

Proof We have established (1.53) which follows:

$$|(L_n(f))(s) - f(s)| \leq |f(s)| \, |(L_n(1))(s) - 1| + K((L_n(|\cdot - s|^{\alpha}))(s)). \quad (1.68)$$

Consider $x \in H : \|x\| = 1$. Then

$$|\langle ((L_n(f))(A) - f(A))x, x \rangle| = \left| \int_{m-0}^{M} ((L_n(f))(s) - f(s)) \, d\langle E_s x, x \rangle \right| \leq \quad (1.69)$$

$$\int_{m-0}^{M} |(L_n(f))(s) - f(s)| \, d\langle E_s x, x \rangle \overset{(1.68)}{\leq}$$

$$\int_{m-0}^{M} |f(s)| \, |(L_n(1))(s) - 1| \, d\langle E_s x, x \rangle + K \int_{m-0}^{M} \left(\left(L_n \left(|\cdot - s|^{\alpha} \right) \right)(s) \right) d\langle E_s x, x \rangle =$$

$$\langle (|f(A)| \, |(L_n(1))(A) - 1_H|) x, x \rangle + K \left\langle \left(\left(L_n \left(|\cdot - A|^{\alpha} \right) \right)(A) \right) x, x \right\rangle \leq$$

$$\|f(A)\| \, \|(L_n(1))(A) - 1_H\| + K \left\| \left(L_n \left(|\cdot - A|^{\alpha} \right) \right)(A) \right\|, \quad (1.70)$$

proving the claim. ■

We continue with

Theorem 1.10 *Let* $\{L_N\}_{N \in \mathbb{N}}$ *be a sequence of positive linear operators from* $C([a,b])$ *into itself. Let* $f : [a,b] \to \mathbb{R}$ *be such that* $f^{(n)} \in C([a,b])$ *and*

$$\left| f^{(n)}(z) - f^{(n)}(s) \right| \leq K \, |z - s|^{\alpha}, \; K > 0, \quad (1.71)$$

$0 < \alpha \leq 1$, $\forall \, z, s \in [a,b]$.
Then it holds

$$\|(L_N(f))(A) - f(A)\| \leq \|f(A)\| \, \|(L_N(1))(A) - 1_H\| +$$

$$\sum_{k=1}^{n} \frac{1}{k!} \left\| f^{(k)}(A) \right\| \left\| \left(L_N \left((\cdot - A)^k \right) \right)(A) \right\| + \frac{K}{\prod_{i=1}^{n} (i + \alpha)} \| \left(L_N \left(|\cdot - A|^{n+\alpha} \right) \right)(A) \|, \quad (1.72)$$

$\forall \, N \in \mathbb{N}$.
Assuming further that

$$\|L_N(1)\|_{\infty} \leq \mu, \; \forall \, N \in \mathbb{N}; \; \mu > 0, \quad (1.73)$$

and $(L_N(1))(A) \to 1_H$, $\left(L_N\left(|\cdot - A|^{n+\alpha}\right)\right)(A) \to 0_H$, *uniformly, as* $N \to \infty$, *we get that* $(L_N(f))(A) \to f(A)$, *uniformly, as* $N \to \infty$, $\forall\, f \in C^n([a,b])$, *fulfilling (1.71).*

Proof Here $f : [a,b] \to \mathbb{R}$ is such that $f^{(n)} \in C([a,b])$. Let $s \in [a,b]$, $n \in \mathbb{N}$. Then

$$f(t) = \sum_{k=0}^{n} \frac{f^{(k)}(s)}{k!}(t-s)^k + R_n(t,s), \tag{1.74}$$

where

$$R_n(t,s) = \frac{1}{(n-1)!}\int_s^t \left[f^{(n)}(z) - f^{(n)}(s)\right](t-z)^{n-1}\,dz, \tag{1.75}$$

$\forall\, t, s \in [a,b]$.

Under the assumption (1.71), next we estimate $R_n(t,s)$:

Let $t \geq s$, then

$$\left|\int_s^t \left[f^{(n)}(z) - f^{(n)}(s)\right](t-z)^{n-1}\,dz\right| \leq \int_s^t \left|f^{(n)}(z) - f^{(n)}(s)\right|(t-z)^{n-1}\,dz \leq$$

$$K\int_s^t |z-s|^{\alpha}(t-z)^{n-1}\,dz = K\int_s^t (t-z)^{n-1}(z-s)^{(\alpha+1)-1}\,dz = \tag{1.76}$$

$$K\frac{\Gamma(n)\,\Gamma(\alpha+1)}{\Gamma(n+\alpha+1)}(t-s)^{n+\alpha}.$$

So, when $t \geq s$ we get

$$\left|\int_s^t \left[f^{(n)}(z) - f^{(n)}(s)\right](t-z)^{n-1}\,dz\right| \leq K\frac{\Gamma(n)\,\Gamma(\alpha+1)}{\Gamma(n+\alpha+1)}(t-s)^{n+\alpha}. \tag{1.77}$$

Let $t \leq s$, then

$$\left|\int_s^t \left[f^{(n)}(z) - f^{(n)}(s)\right](t-z)^{n-1}\,dz\right| = \left|\int_t^s \left[f^{(n)}(z) - f^{(n)}(s)\right](z-t)^{n-1}\,dz\right|$$

$$\leq \int_t^s \left|f^{(n)}(z) - f^{(n)}(s)\right|(z-t)^{n-1}\,dz \leq K\int_t^s |z-s|^{\alpha}(z-t)^{n-1}\,dz \tag{1.78}$$

$$= K\int_t^s (s-z)^{(\alpha+1)-1}(z-t)^{n-1}\,dz = K\frac{\Gamma(\alpha+1)\,\Gamma(n)}{\Gamma(n+1+\alpha)}(s-t)^{n+\alpha}.$$

We have proved that

$$\left| \int_s^t \left[f^{(n)}(z) - f^{(n)}(s) \right] (t-z)^{n-1} dz \right| \leq K \frac{\Gamma(n)\Gamma(\alpha+1)}{\Gamma(n+\alpha+1)} |t-s|^{n+\alpha} \quad (1.79)$$

$$= K \frac{(n-1)!}{(1+\alpha)(2+\alpha)\dots(n+\alpha)} |t-s|^{n+\alpha}, \ \forall\, t,s \in [a,b].$$

Hence it holds

$$|R_n(t,s)| = \frac{1}{(n-1)!} \left| \int_s^t \left[f^{(n)}(z) - f^{(n)}(s) \right] (t-z)^{n-1} dz \right| \leq$$

$$\frac{K}{(1+\alpha)(2+\alpha)\dots(n+\alpha)} |t-s|^{n+\alpha}, \ \forall\, t,s \in [a,b]. \quad (1.80)$$

Let now L_N, $N \in \mathbb{N}$, be a sequence of positive linear operators from $C([a,b])$ into itself.

Then we get

$$|(L_N(R_n(\cdot,s)))(s)| \leq (L_N(|(R_n(\cdot,s))|))(s) \leq \quad (1.81)$$

$$\frac{K}{\prod_{i=1}^{n}(i+\alpha)} \left(L_N\left(|\cdot - s|^{n+\alpha} \right) \right)(s), \ \forall\, s \in [a,b].$$

Above, $\left(L_N\left(|\cdot - s|^{n+\alpha} \right) \right)(s)$ is continuous in $s \in [a,b]$ (by Lemma 1.4). We can rewrite (1.74) as follows

$$f(\cdot) - f(s) = \sum_{k=1}^{n} \frac{f^{(k)}(s)}{k!} (\cdot - s)^k + R_n(\cdot,s), \quad (1.82)$$

and we notice that $R_n(\cdot,s) \in C([a,b])$, here we keep s fixed.

Hence we find

$$(L_N(f))(s) - f(s)(L_N(1))(s) =$$

$$\sum_{k=1}^{n} \frac{f^{(k)}(s)}{k!} \left(L_N(\cdot - s)^k \right)(s) + (L_N(R_n(\cdot,s)))(s), \ \forall\, s \in [a,b]. \quad (1.83)$$

Therefore we have

$$(L_N(f))(s) - f(s) = (L_N(f))(s) - f(s) - f(s)(L_N(1))(s) + f(s)(L_N(1))(s) =$$

$$(L_N(f))(s) - f(s)(L_N(1))(s) + f(s)((L_N(1))(s) - 1) =$$

$$f(s)((L_N(1))(s) - 1) + \sum_{k=1}^{n} \frac{f^{(k)}(s)}{k!} \left(L_N(\cdot - s)^k\right)(s) + (L_N(R_n(\cdot, s)))(s), \quad (1.84)$$

$\forall\, s \in [a, b]$.

Thus, it holds

$$(L_N(f))(s) - f(s) = f(s)((L_N(1))(s) - 1) + \sum_{k=1}^{n} \frac{f^{(k)}(s)}{k!} \left(L_N(\cdot - s)^k\right)(s)$$
$$+ (L_N(R_n(\cdot, s)))(s), \ \forall\, s \in [a, b]. \quad (1.85)$$

Above, $\left(L_N(\cdot - s)^k\right)(s)$ is continuous in $s \in [a, b]$, for all $k = 1, ..., n$.

Furthermore it holds

$$|(L_N(f))(s) - f(s)| \overset{(1.81)}{\leq} |f(s)| \, |(L_N(1))(s) - 1| +$$

$$\sum_{k=1}^{n} \frac{\left|f^{(k)}(s)\right|}{k!} \left|\left(L_N\left((\cdot - s)^k\right)\right)(s)\right| + \frac{K}{\prod_{i=1}^{n}(i + \alpha)} \left(L_N\left(|\cdot - s|^{n+\alpha}\right)\right)(s), \quad (1.86)$$

$\forall\, s \in [a, b]$.

Next we observe that

$$\|(L_N(f))(A) - f(A)\| = \||(L_N(f))(A) - f(A)|\| = \quad (1.87)$$

$$\sup_{x \in H: \|x\|=1} |\langle |(L_N(f))(A) - f(A)| \, x, x\rangle| =$$

$$\sup_{x \in H: \|x\|=1} \left| \int_{m-0}^{M} |(L_N(f))(s) - f(s)| \, d\langle E_s x, x\rangle \right| = \quad (1.88)$$

$$\sup_{x\in H:\|x\|=1}\int_{m-0}^{M}|(L_N(f))(s)-f(s)|\,d\langle E_s x,x\rangle \leq$$

$$\sup_{x\in H:\|x\|=1}\langle |f(A)|\,|(L_n(1))A-1_H|\,x,x\rangle +$$

$$\sum_{k=1}^{n}\frac{1}{k!}\sup_{x\in H:\|x\|=1}\left\langle \left|f^{(k)}(A)\right|\left|\left(L_n\left((\cdot-A)^k\right)\right)(A)\right|x,x\right\rangle +$$

$$\frac{K}{\prod_{i=1}^{n}(i+\alpha)}\sup_{x\in H:\|x\|=1}\left\langle \left(L_n\left(|\cdot-A|^{n+\alpha}\right)\right)(A)\,x,x\right\rangle =$$

$$\||f(A)||\,(L_n(1))(A)-1_H\||+\sum_{k=1}^{n}\frac{1}{k!}\left\|\left|f^{(k)}(A)\right|\left|\left(L_N\left((\cdot-A)^k\right)\right)(A)\right|\right\|+$$

$$\frac{K}{\prod_{i=1}^{n}(i+\alpha)}\left\|\left(L_n\left(|\cdot-A|^{n+\alpha}\right)\right)(A)\right\| \leq \tag{1.89}$$

$$\|f(A)\|\||(L_n(1))(A)-1_H\|+\sum_{k=1}^{n}\frac{1}{k!}\left\|f^{(k)}(A)\right\|\left\|\left(L_N\left((\cdot-A)^k\right)\right)(A)\right\|+$$

$$\frac{K}{\prod_{i=1}^{n}(i+\alpha)}\left\|\left(L_n\left(|\cdot-A|^{n+\alpha}\right)\right)(A)\right\|,$$

proving the inequality (1.72).

By (1.37) we have ($k=1,...,n;\ 0<\alpha\leq 1$) that

$$\left|\left(L_N(\cdot-s)^k\right)(s)\right| \leq \left((L_N(1))(s)\right)^{\left(\frac{n+\alpha-k}{n+\alpha}\right)}\left(\left(L_N\left(|\cdot-s|^{n+\alpha}\right)\right)(s)\right)^{\frac{k}{n+\alpha}}, \tag{1.90}$$

$\forall\, s\in[a,b]$, $\forall\, N\in\mathbb{N}$.

By assumption (1.73) we get

$$\left|\left(L_N(\cdot-s)^k\right)(s)\right| \leq \mu^{\left(\frac{n+\alpha-k}{n+\alpha}\right)}\left(\left(L_N\left(|\cdot-s|^{n+\alpha}\right)\right)(s)\right)^{\frac{k}{n+\alpha}}, \tag{1.91}$$

$\forall\, s\in[a,b]$, $\forall\, N\in\mathbb{N}$; $k=1,...,n$; $0<\alpha\leq 1$.

Hence we derive

$$\left\|\left(L_N\left((\cdot-A)^k\right)\right)(A)\right\| = \left\|\left|L_N\left((\cdot-A)^k\right)(A)\right|\right\| =$$

$$\sup_{x\in H:\|x\|=1}\left|\left\langle\left|L_N\left((\cdot-A)^k\right)(A)\right|x,x\right\rangle\right| =$$

$$\sup_{x \in H: \|x\|=1} \left\langle \left| L_N \left((\cdot - A)^k \right) (A) \right| x, x \right\rangle = \tag{1.92}$$

$$\sup_{x \in H: \|x\|=1} \int_{m-0}^{M} \left| \left(L_N \left((\cdot - s)^k \right) \right) (s) \right| d \langle E_s x, x \rangle \leq$$

$$\mu^{\left(\frac{n+\alpha-k}{n+\alpha} \right)} \sup_{x \in H: \|x\|=1} \int_{m-0}^{M} \left(\left(L_N \left(|\cdot - s|^{n+\alpha} \right) \right) (s) \right)^{\frac{k}{n+\alpha}} d \langle E_s x, x \rangle =$$

$$\mu^{\left(\frac{n+\alpha-k}{n+\alpha} \right)} \sup_{x \in H: \|x\|=1} \left\langle \left(\left(L_N \left(|\cdot - A|^{n+\alpha} \right) \right) (A) \right)^{\frac{k}{n+\alpha}} x, x \right\rangle$$

(by Hölder-McCarthy's inequality (1.12))

$$\leq \mu^{\left(\frac{n+\alpha-k}{n+\alpha} \right)} \sup_{x \in H: \|x\|=1} \left(\left\langle \left(\left(L_N \left(|\cdot - A|^{n+\alpha} \right) \right) (A) \right) x, x \right\rangle \right)^{\frac{k}{n+\alpha}} = \tag{1.93}$$

$$\mu^{\left(\frac{n+\alpha-k}{n+\alpha} \right)} \left(\sup_{x \in H: \|x\|=1} \left| \left\langle \left(\left(L_N \left(|\cdot - A|^{n+\alpha} \right) \right) (A) \right) x, x \right\rangle \right| \right)^{\frac{k}{n+\alpha}} =$$

$$\mu^{\left(\frac{n+\alpha-k}{n+\alpha} \right)} \left\| \left(L_N \left(|\cdot - A|^{n+\alpha} \right) \right) (A) \right\|^{\frac{k}{n+\alpha}}.$$

We have proved that

$$\left\| \left(L_N \left((\cdot - A)^k \right) \right) (A) \right\| \leq \mu^{\left(\frac{n+\alpha-k}{n+\alpha} \right)} \left\| \left(L_N \left(|\cdot - A|^{n+\alpha} \right) \right) (A) \right\|^{\frac{k}{n+\alpha}}, \tag{1.94}$$

all $k = 1, ..., n$, $0 < \alpha \leq 1$, $\forall\, N \in \mathbb{N}$.

By (1.94) and by assuming that $(L_N(1))(A) \to 1_H$ and $\left(L_N \left(|\cdot - A|^{n+\alpha} \right) \right)(A) \to 0_H$, uniformly, as $N \to \infty$, we get that $\left(L_N \left((\cdot - A)^k \right) \right)(A) \to 0_H$, and $(L_N(f))(A) \to f(A)$, uniformly, $\forall\, f \in C([a,b])$, under assumptions (1.71), (1.73). ■

We continue with

Theorem 1.11 *Let $\{L_N\}_{N \in \mathbb{N}}$ be a sequence of positive linear operators from $C([a,b])$ into itself. Let $f : [a,b] \to \mathbb{R}$ be such that $f^{(n)} \in C([a,b])$ and*

$$\left| f^{(n)}(z) - f^{(n)}(s) \right| \leq K |z - s|^{\alpha}, \; K > 0, \tag{1.95}$$

$0 < \alpha \leq 1$, $\forall\, z, s \in [a,b]$.

Then it holds

$$\left\| (L_N(f))(A) - f(A) - \sum_{k=1}^{n} \frac{f^{(k)}(A)}{k!} \left(L_N(\cdot - A)^k\right)(A) \right\| \leq$$

$$\|f(A)\| \, \|(L_N(1))(A) - 1_H\| + \frac{K}{\prod_{i=1}^{n}(i+\alpha)} \| \left(L_N\left(|\cdot - A|^{n+\alpha}\right)\right)(A) \|. \quad (1.96)$$

Conclusion: If $(L_N(1))(A) \to 1_H$ and $\left(L_N\left(|\cdot - A|^{n+\alpha}\right)\right)(A) \to 0_H$, uniformly, as $N \to \infty$, then

$$\left[(L_N(f))(A) - \sum_{k=1}^{n} \frac{f^{(k)}(A)}{k!} \left(L_N(\cdot - A)^k\right)(A) \right] \to f(A),$$

uniformly, as $N \to \infty$, $\forall\, f$ as above.

Proof The next is a continuous function in $s \in [a, b]$:

$$(L_N(f))(s) - f(s) - \sum_{k=1}^{n} \frac{f^{(k)}(s)}{k!} \left(L_N(\cdot - s)^k\right)(s) \overset{(1.85)}{=}$$

$$f(s)\left((L_N(1))(s) - 1\right) + \left(L_N(R_n(\cdot, s))(s)\right), \ \forall\, s \in [a, b]. \quad (1.97)$$

Hence it holds

$$\left\| (L_N(f))(A) - f(A) - \sum_{k=1}^{n} \frac{f^{(k)}(A)}{k!} \left(L_N(\cdot - A)^k\right)(A) \right\| =$$

$$\left\| \left| (L_N(f))(A) - f(A) - \sum_{k=1}^{n} \frac{f^{(k)}(A)}{k!} \left(L_N(\cdot - A)^k\right)(A) \right| \right\| =$$

$$\sup_{x \in H: \|x\|=1} \left| \left\langle \left| (L_N(f))(A) - f(A) - \sum_{k=1}^{n} \frac{f^{(k)}(A)}{k!} \left(L_N(\cdot - A)^k\right)(A) \right| x, x \right\rangle \right| =$$

$$\sup_{x \in H: \|x\|=1} \left\langle \left| (L_N(f))(A) - f(A) - \sum_{k=1}^{n} \frac{f^{(k)}(A)}{k!} \left(L_N(\cdot - A)^k\right)(A) \right| x, x \right\rangle =$$

$$(1.98)$$

$$\sup_{x \in H: \|x\|=1} \int_{m-0}^{M} \left| f(s)\left((L_N(1))(s) - 1\right) + \left(L_N(R_n(\cdot, s))\right)(s) \right| d\langle E_s x, x \rangle \leq$$

$$\sup_{x\in H:\|x\|=1}\int_{m-0}^{M}|f(s)|\,|(L_N(1))(s)-1|\,d\langle E_s x,x\rangle+ \tag{1.99}$$

$$\sup_{x\in H:\|x\|=1}\int_{m-0}^{M}(L_N(|R_n(\cdot,s)|))(s)\,d\langle E_s x,x\rangle\overset{(1.81)}{\leq}$$

$$\sup_{x\in H:\|x\|=1}\langle|f(A)|\,|(L_N(1))(A)-1_H|\,x,x\rangle+$$

$$\sup_{x\in H:\|x\|=1}\frac{K}{\prod_{i=1}^{n}(i+\alpha)}\int_{m-0}^{M}\left(L_N\left(|\cdot-s|^{n+\alpha}\right)\right)(s)\,d\langle E_s x,x\rangle\overset{(1.13)}{\leq}$$

$$\||f(A)|\|\,\||(L_N(1))(A)-1_H|\|+\frac{K}{\prod_{i=1}^{n}(i+\alpha)}\left\|\left(L_N\left(|\cdot-A|^{n+\alpha}\right)\right)(A)\right\|=$$

$$\|f(A)\|\,\|(L_N(1))(A)-1_H\|+\frac{K}{\prod_{i=1}^{n}(i+\alpha)}\|\left(L_N\left(|\cdot-A|^{n+\alpha}\right)\right)(A)\|, \tag{1.100}$$

proving the claim. ■

1.4 Applications

For the next, see [3], pp. 169–170.

Let $g\in C([0,1])$, $N\in\mathbb{N}$, the Nth basic Bernstein polynomial for g is defined by

$$(\beta_N(g))(z)=\sum_{k=0}^{N}\binom{N}{k}g\left(\frac{k}{N}\right)z^k(1-z)^{N-k},\ \forall z\in[0,1]. \tag{1.101}$$

It has the properties:

$$\beta_N(1)=1,\ (\beta_N(id))(z)=z,$$

$$(\beta_N((\cdot-z)))(z)=0,\ \left(\beta_N\left(id^2\right)\right)(z)=\left(1-\frac{1}{N}\right)z^2+\frac{1}{N}z, \tag{1.102}$$

and

$$\left(\beta_N\left((\cdot-z)^2\right)\right)(z)=\frac{z(1-z)}{N},\ \forall z\in[0,1].$$

Here we consider $f\in C([a,b])$, and the general Bernstein positive linear polynomial operators from $C([a,b])$ into itself, defined by (see [10], p. 80)

$$(B_N f)(s) = \sum_{i=0}^{N} \binom{N}{i} f\left(a + i\frac{(b-a)}{N}\right)\left(\frac{s-a}{b-a}\right)^i \left(\frac{b-s}{b-a}\right)^{N-i}, \forall s \in [a,b]. \tag{1.103}$$

By [10], p. 81, we get that

$$\|B_N f - f\|_\infty \leq \frac{5}{4}\omega_1\left(f; \frac{b-a}{\sqrt{N}}\right), \tag{1.104}$$

i.e. $B_N f \to f$, uniformly, as $N \to \infty$, $\forall f \in C([a,b])$, the convergence is given with rates.

We easily obtain that

$$(B_N(1))(s) = 1, \forall s \in [a,b], \text{ i.e. } B_N(1) = 1. \tag{1.105}$$

We notice that

$$\left(\frac{s-a}{b-a}\right) + \left(\frac{b-s}{b-a}\right) = 1, \forall s \in [a,b], \tag{1.106}$$

calling $y := \frac{s-a}{b-a}$, we have $\frac{b-s}{b-a} = 1 - y$.

So we can write

$$(B_N f)(s) = \sum_{i=0}^{N} \binom{N}{i} f\left(a + \frac{i(b-a)}{N}\right) y^i (1-y)^{N-i}. \tag{1.107}$$

We observe that

$$(B_N(id))(s) = \sum_{i=0}^{N} \binom{N}{i} \left(a + \frac{i(b-a)}{N}\right) y^i (1-y)^{N-i}$$

$$= a + (b-a)\sum_{i=0}^{N} \binom{N}{i} \left(\frac{i}{N}\right) y^i (1-y)^{N-i}$$

$$\overset{(1.102)}{=} a + (b-a)y = a + s - a = s,$$

proving $(B_n(id))(s) = s$, i.e. it holds

$$B_N(id) = id. \tag{1.108}$$

We see that

$$(B_N((id - s)))(s) = (B_N(id))(s) - s(B_N(1))(s) \overset{(1.105)}{=} s - s = 0,$$

i.e.

$$(B_N((\cdot - s)))(s) = 0, \forall s \in [a, b]. \tag{1.109}$$

Next we calculate

$$\left(B_N\left(id^2\right)\right)(s) = \sum_{i=0}^{N} \binom{N}{i} \left(a + \frac{i(b-a)}{N}\right)^2 y^i (1-y)^{N-i}$$

$$= \sum_{i=0}^{N} \binom{N}{i} \left(a^2 + 2a(b-a)\frac{i}{N} + (b-a)^2\left(\frac{i}{N}\right)^2\right) y^i (1-y)^{N-i}$$

$$= a^2 + 2a(b-a) \sum_{i=0}^{N} \binom{N}{i} \left(\frac{i}{N}\right) y^i (1-y)^{N-i}$$

$$+ (b-a)^2 \sum_{i=0}^{N} \binom{N}{i} \left(\frac{i}{N}\right)^2 y^i (1-y)^{N-i} \tag{1.110}$$

$$\stackrel{(1.102)}{=} a^2 + 2a(b-a)y + (b-a)^2 \left[\left(1 - \frac{1}{N}\right) y^2 + \frac{1}{N} y\right]$$

$$= a^2 + 2a(s-a) + (b-a)^2 \left[\left(1 - \frac{1}{N}\right) \frac{(s-a)^2}{(b-a)^2} + \frac{1}{N}\left(\frac{s-a}{b-a}\right)\right]$$

$$= a^2 + 2a(s-a) + \left(1 - \frac{1}{N}\right)(s-a)^2 + \frac{1}{N}(b-a)(s-a)$$

$$= a^2 + (s-a)\left(2a + \frac{b-a}{N}\right) + \left(1 - \frac{1}{N}\right)(s-a)^2.$$

We have proved that

$$\left(B_N\left(id^2\right)\right)(s) = a^2 + (s-a)\left(2a + \frac{b-a}{N}\right) + \left(1 - \frac{1}{N}\right)(s-a)^2, \forall s \in [a, b]. \tag{1.111}$$

Finally we calculate

$$\left(B_N\left((id - s)^2\right)\right)(s) = \left(B_N\left(\left(id^2 - 2s\, id + s^2\right)\right)\right)(s) =$$

$$\left(B_N\left(id^2\right)\right)(s) - 2s\left(B_N(id)\right)(s) + s^2\left(B_N(1)\right)(s) =$$

$$a^2 + (s-a)\left(2a + \frac{b-a}{N}\right) + \left(1 - \frac{1}{N}\right)(s-a)^2 - 2s^2 + s^2 = \tag{1.112}$$

$$a^2+(s-a)\left(2a+\frac{b-a}{N}\right)+\left(1-\frac{1}{N}\right)(s-a)^2-s^2=$$

$$2a^2+(s-a)\left(2a+\frac{b-a}{N}\right)-2as-\frac{1}{N}(s-a)^2.$$

We have proved that

$$\left(B_N\left((id-s)^2\right)\right)(s)=2a^2+(s-a)\left(2a+\frac{b-a}{N}\right)-2as-\frac{1}{N}(s-a)^2, \tag{1.113}$$

$\forall\, s\in[a,b]$.

Notice that

$$\lim_{N\to\infty}\left(B_N\left((id-s)^2\right)\right)(s)=0, \tag{1.114}$$

as well as

$$\lim_{N\to\infty}\left(B_N\left(id^2\right)\right)(s)=s^2,\ \forall s\in[a,b], \tag{1.115}$$

both uniformly.

Next we apply the results of Sect. 1.3 for the case of B_N operators. Here again $Sp(A)\subseteq[m,M]\subset(a,b)$; A as selfadjoint operator on the Hilbert space H. By (1.1) and (1.2) and $x\in H:\|x\|=1$, we get $\int_{m-0}^{M}d\langle E_\lambda x,x\rangle=1$.

So Theorem 1.7 is going to read as follows:

Corollary 1.12 *Let $f\in C([a,b])$. Then*

$$\|(B_N(f))(A)-f(A)\|\leq\|B_Nf-f\|_{\infty,[a,b]},\ \forall\, N\in\mathbb{N}. \tag{1.116}$$

By earlier comments on B_N, see (1.104), we get that $\lim_{N\to\infty}(B_N(f))(A)=f(A)$, uniformly, $\forall\, f\in C([a,b])$.

Next we apply Theorem 1.8 to B_N operators

Corollary 1.13 *Let $f:[a,b]\to\mathbb{R}$. Assume that*

$$|f(t)-f(s)|\leq K|t-s|^{\alpha},\ \forall\, t,s\in[a,b], \tag{1.117}$$

where $0<\alpha\leq1$, $K>0$.

Then

$$\|(B_N(f))(A)-f(A)\|\leq K\left\|\left(B_N\left(id^2\right)\right)(A)-A^2\right\|^{\frac{\alpha}{2}},\ \forall\, N\in\mathbb{N}. \tag{1.118}$$

Since (see Remark 1.14 next) $\left(B_N\left(id^2\right)\right)(A)\to A^2$, uniformly, as $N\to\infty$, we get that $(B_N(f))(A)\to f(A)$, uniformly, as $N\to\infty$, $\forall\, f\in C([a,b])$, fulfilling (1.117).

Remark 1.14 Indeed it holds

$$\left\| \left(B_N \left(id^2\right)\right)(A) - A^2 \right\| = \sup_{\substack{x \in H: \\ \|x\|=1}} \left| \left\langle \left(\left(B_N \left(id^2\right)\right)(A) - A^2\right) x, x \right\rangle \right| =$$

$$\left| \int_{m-0}^{M} \left(\left(B_N \left(id^2\right)\right)(s) - s^2\right) d \langle E_s x, x \rangle \right| \leq \int_{m-0}^{M} \left| \left(B_N \left(id^2\right)\right)(s) - s^2 \right| d \langle E_s x, x \rangle$$

$$\leq \left\| \left(B_N \left(id^2\right)\right)(s) - s^2 \right\|_{\infty, [a,b]} \to 0, \tag{1.119}$$

as $N \to \infty$, by (1.115).

Proving that $\left(B_N \left(id^2\right)\right)(A) \to A^2$, uniformly, as $N \to \infty$.

Corollary 1.15 (to Theorem 1.9) *Let* $f : [a, b] \to \mathbb{R}$:

$$|f(t) - f(s)| \leq K |t - s|^{\alpha}, \forall\, t, s \in [a, b], \tag{1.120}$$

where $0 < \alpha \leq 1$, $K > 0$.

Then

$$\|(B_N(f))(A) - f(A)\| \leq K \|(B_N(|\cdot - A|^{\alpha}))(A)\|, \forall\, N \in \mathbb{N}. \tag{1.121}$$

Since $(B_N(|\cdot - A|^{\alpha}))(A) \to 0_H$, uniformly, (see Remarks 1.16 and 1.18) then $(B_N(f))(A) \to f(A)$, uniformly, as $N \to \infty$, for every f as above, see (1.120).

Remark 1.16 We easily obtain (by Hölder's inequality and (1.105))

$$(B_N(|\cdot - s|^{\alpha}))(s) \leq \left(\left(B_N\left((\cdot - s)^2\right)\right)(s)\right)^{\frac{\alpha}{2}}, \forall\, s \in [a, b]. \tag{1.122}$$

Hence it holds

$$\|(B_N(|\cdot - A|^{\alpha}))(A)\| = \sup_{\substack{x \in H: \\ \|x\|=1}} |\langle ((B_N(|\cdot - A|^{\alpha}))(A)) x, x \rangle| =$$

$$\sup_{\substack{x \in H: \\ \|x\|=1}} \langle ((B_N(|\cdot - A|^{\alpha}))(A)) x, x \rangle = \sup_{\substack{x \in H: \\ \|x\|=1}} \int_{m-0}^{M} (B_N(|\cdot - s|^{\alpha}))(s)\, d \langle E_s x, x \rangle \overset{(1.122)}{\leq}$$

$$\sup_{\substack{x \in H: \\ \|x\|=1}} \int_{m-0}^{M} \left(\left(B_N\left(|\cdot - s|^2\right)\right)(s)\right)^{\frac{\alpha}{2}} d \langle E_s x, x \rangle = \tag{1.123}$$

$$\sup_{\substack{x \in H: \\ \|x\|=1}} \left\langle \left(\left(B_N\left(|\cdot - A|^2\right)\right)(A)\right)^{\frac{\alpha}{2}} x, x \right\rangle \overset{(1.12)}{\leq}$$

$$\sup_{\substack{x \in H: \\ \|x\|=1}} \left(\left\langle \left(\left(B_N \left(|\cdot - A|^2 \right) \right)(A) \right) x, x \right\rangle \right)^{\frac{\alpha}{2}} = \tag{1.124}$$

$$\sup_{\substack{x \in H: \\ \|x\|=1}} \left| \left\langle \left(\left(B_N \left(|\cdot - A|^2 \right) \right)(A) \right) x, x \right\rangle \right|^{\frac{\alpha}{2}} = \left\| \left(B_N \left((\cdot - A)^2 \right) \right)(A) \right\|^{\frac{\alpha}{2}}.$$

That is we have

$$\|(B_N(|\cdot - A|^{\alpha}))(A)\| \leq \left\| \left(B_N \left((\cdot - A)^2 \right) \right)(A) \right\|^{\frac{\alpha}{2}}. \tag{1.125}$$

Corollary 1.17 (to Theorem 1.10 *for* $n = 1$, $\alpha = 1$) *Let* $f : [a, b] \to \mathbb{R}$:

$$\left| f'(z) - f'(s) \right| \leq K |z - s|, \; K > 0, \; \forall\, z, s \in [a, b]. \tag{1.126}$$

Then

$$\|(B_N(f))(A) - f(A)\| \leq \frac{K}{2} \left\| \left(B_N \left((\cdot - A)^2 \right) \right)(A) \right\|, \; \forall\, N \in \mathbb{N}. \tag{1.127}$$

Proof See also (1.105) and (1.109). ■

We make

Remark 1.18 We observe the following:

$$\left\| \left(B_N \left((\cdot - A)^2 \right) \right)(A) \right\| = \sup_{\substack{x \in H: \\ \|x\|=1}} \left| \left\langle \left(\left(B_N \left((\cdot - A)^2 \right) \right)(A) \right) x, x \right\rangle \right| =$$

$$\sup_{\substack{x \in H: \\ \|x\|=1}} \left\langle \left(\left(B_N \left((\cdot - A)^2 \right) \right)(A) \right) x, x \right\rangle = \sup_{\substack{x \in H: \\ \|x\|=1}} \int_{m-0}^{M} \left(B_N \left((\cdot - s)^2 \right) \right)(s)\, d \langle E_s x, x \rangle \leq$$

$$\left\| \left(B_N \left((\cdot - s)^2 \right) \right)(s) \right\|_{\infty, [a,b]} \to 0, \tag{1.128}$$

as $N \to \infty$, (by (1.114)). That is proving

$$\left\| \left(B_N \left((\cdot - A)^2 \right) \right)(A) \right\| \to 0,$$

and by (1.127), we derive that $(B_N(f))(A) \to f(A)$, uniformly, as $N \to \infty$, for every f as in (1.126).

References

1. G.A. Anastassiou, *Moments in Probability and Approximation Theory*, Longman Scientific & Technical, Pitman Research Notes in Mathematics Series, vol. 287, Wiley, Essex, New York (1993)
2. G. Anastassiou, *Self Adjoint Operator Korovkin type Quantitative Approximations*, Acta Mathematica Universitatis Comenianae, accepted (2016)
3. R.G. Bartle, *The Elements of Real Analysis*, 2nd edn. (Wiley, New York, 1976)
4. S.S. Dragomir, *Inequalities for Functions of Selfadjoint Operators on Hilbert Spaces* (2011). ajmaa.org/RGMIA/monographs/InFuncOp.pdf
5. S. Dragomir, *Operator Inequalities of Ostrowski and Trapezoidal Type* (Springer, New York, 2012)
6. G. Helmberg, *Introduction to Spectral Theory in Hilbert Space* (John Wiley & Sons Inc., New York, 1969)
7. T. Furuta, J. Mićić Hot, J. Pečarić, Y. Seo, *Mond-Pečarić Method in Operator Inequalities. Inequalities for Bounded Selfadjoint Operators on a Hilbert Space*, Element, Zagreb (2005)
8. C.A. McCarthy, c_p. Israel J. Math. **5**, 249–271 (1967)
9. O. Shisha, B. Mond, The degree of convergence of sequences of linear positive operators. Nat. Acad. of Sci. U.S. **60**, 1196–1200 (1968)
10. L. Shumaker, *Spline Functions Basic Theory* (Wiley-Interscience, New York, 1981)

Chapter 2
Self Adjoint Operator Korovkin and Polynomial Direct Approximations with Rates

Here we present self adjoint operator Korovkin type theorems, via self adjoint operator Shisha–Mond type inequalities, also we give self adjoint operator polynomial approximations. This is a quantitative treatment to determine the degree of self adjoint operator uniform approximation with rates, of sequences of self adjoint operator positive linear operators. The same kind of work is performed over important operator polynomial sequences. Our approach is direct based on Gelfand isometry. It follows [1].

2.1 Background

Let A be a selfadjoint linear operator on a complex Hilbert space $(H; \langle \cdot, \cdot \rangle)$. The Gelfand map establishes a $*-$isometrically isomorphism Φ between the set $C(Sp(A))$ of all continuous functions defined on the spectrum of A, denoted $Sp(A)$, and the C^*-algebra $C^*(A)$ generated by A and the identity operator 1_H on H as follows (see e.g. [2, p. 3]):

For any $f, g \in C(Sp(A))$ and any $\alpha, \beta \in \mathbb{C}$ we have

(i) $\Phi(\alpha f + \beta g) = \alpha \Phi(f) + \beta \Phi(g)$;

(ii) $\Phi(fg) = \Phi(f)\Phi(g)$ (the operation composition is on the right) and $\Phi(\overline{f}) = (\Phi(f))^*$;

(iii) $\|\Phi(f)\| = \|f\| := \sup_{t \in Sp(A)} |f(t)|$;

(iv) $\Phi(f_0) = 1_H$ and $\Phi(f_1) = A$, where $f_0(t) = 1$ and $f_1(t) = t$, for $t \in Sp(A)$.

G.A. Anastassiou, *Intelligent Comparisons II: Operator Inequalities and Approximations*, Studies in Computational Intelligence 699,
DOI 10.1007/978-3-319-51475-8_2

With this notation we define

$$f(A) := \Phi(f), \text{for all } f \in C(Sp(A)),$$

and we call it the continuous functional calculus for a selfadjoint operator A.

If A is a selfadjoint operator and f is a real valued continuous function on $Sp(A)$ then $f(t) \geq 0$ for any $t \in Sp(A)$ implies that $f(A) \geq 0$, i.e. $f(A)$ is a positive operator on H. Moreover, if both f and g are real valued continuous functions on $Sp(A)$ then the following important property holds:

(P) $f(t) \geq g(t)$ for any $t \in Sp(A)$, implies that $f(A) \geq g(A)$ in the operator order of $B(H)$.

Equivalently, we use (see [3], pp. 7–8).

Let U be a selfadjoint operator on the complex Hilbert space $(H, \langle \cdot, \cdot \rangle)$ with the spectrum $Sp(U)$ included in the interval $[m, M]$ for some real numbers $m < M$ and $\{E_\lambda\}_\lambda$ be its spectral family.

Then for any continuous function $f : [a, b] \to \mathbb{C}$, where [m, M] $\subset$ (a, b), it is well known that we have the following spectral representation in terms of the Riemann–Stieljes integral:

$$\langle f(U)x, y \rangle = \int_{m-0}^{M} f(\lambda) d(\langle E_\lambda x, y \rangle),$$

for any $x, y \in H$. The function $g_{x,y}(\lambda) := \langle E_\lambda x, y \rangle$ is of bounded variation on the interval $[m, M]$, and

$$g_{x,y}(m-0) = 0 \text{ and } g_{x,y}(M) = \langle x, y \rangle,$$

for any $x, y \in H$. Furthermore, it is known that $g_x(\lambda) := \langle E_\lambda x, x \rangle$ is increasing and right continuous on $[m, M]$.

In this chapter we will be using a lot the formula

$$\langle f(U)x, x \rangle = \int_{m-0}^{M} f(\lambda) d(\langle E_\lambda x, x \rangle), \ \forall x \in H.$$

As a symbol we can write

$$f(U) = \int_{m-0}^{M} f(\lambda) dE_\lambda.$$

Above, $m = \min\{\lambda | \lambda \in Sp(U)\} := \min Sp(U)$, $M = \max\{\lambda | \lambda \in Sp(U)\} := \max Sp(U)$. The projections $\{E_\lambda\}_{\lambda \in \mathbb{R}}$, are called the spectral family of A, with the properties:

(a) $E_\lambda \leq E_{\lambda'}$ for $\lambda \leq \lambda'$;

(b) $E_{m-0} = 0_H$ (zero operator), $E_M = 1_H$ (identity operator) and $E_{\lambda+0} = E_\lambda$ for all $\lambda \in \mathbb{R}$.

Furthermore

$$E_\lambda := \varphi_\lambda(U), \ \forall \lambda \in \mathbb{R},$$

is a projection which reduces U, with

$$\varphi_\lambda(s) := \begin{cases} 1, \text{ for } -\infty < s \leq \lambda, \\ 0, \text{ for } \lambda < s < +\infty. \end{cases}$$

The spectral family $\{E_\lambda\}_{\lambda \in \mathbb{R}}$ determines uniquely the self-adjoint operator U and vice versa.

For more on the topic see [4], pp. 256–266, and for more detalis see there pp. 157–266. See also [5].

Some more basics are given (we follow [3], pp. 1–5):

Let $(H; \langle \cdot, \cdot \rangle)$ be a Hilbert space over $\mathbb{C}$. A bounded linear operator A defined on H is selfjoint, i.e., $A = A^*$, iff $\langle Ax, x \rangle \in \mathbb{R}, \forall x \in H$, and if A is selfadjoint, then

$$\|A\| = \sup_{x \in H: \|x\|=1} |\langle Ax, x \rangle|.$$

Let A, B be selfadjoint operators on H. Then $A \leq B$ iff $\langle Ax, x \rangle \leq \langle Bx, x \rangle, \forall x \in H$.

In particular, A is called positive if $A \geq 0$.

Denote by

$$\mathcal{P} := \left\{ \varphi(s) := \sum_{k=0}^{n} \alpha_k s^k | n \geq 0, \alpha_k \in \mathbb{C}, 0 \leq k \leq n \right\}.$$

If $A \in \mathcal{B}(H)$ (the Banach algebra of all bounded linear operators defined on H, i.e. from H into itself) is selfadjoint, and $\varphi(s) \in \mathcal{P}$ has real coefficients, then $\varphi(A)$ is selfadjoint, and

$$\|\varphi(A)\| = \max\{|\varphi(\lambda)|, \lambda \in Sp(A)\}.$$

If φ is any function defined on $\mathbb{R}$ we define

$$\|\varphi\|_A := \sup\{|\varphi(\lambda)|, \lambda \in Sp(A)\}.$$

If A is selfadjoint operator on Hilbert space H and φ is continuous and given that $\varphi(A)$ is selfadjoint, then $\|\varphi(A)\| = \|\varphi\|_A$. And if φ is a continuous real valued function so it is $|\varphi|$, then $\varphi(A)$ and $|\varphi|(A) = |\varphi(A)|$ are selfadjoint operators (by [3], p. 4, Theorem 7).

Hence it holds

$$\||\varphi(A)|\| = \||\varphi|\|_A = \sup\{||\varphi(\lambda)||, \lambda \in Sp(A)\}$$

$$= \sup\{|\varphi(\lambda)|, \lambda \in Sp(A)\} = \|\varphi\|_A = \|\varphi(A)\|,$$

that is $\||\varphi(A)|\| = \|\varphi(A)\|$.

For a selfadjoint operator $A \in \mathcal{B}(H)$ which is positive, there exists a unique positive selfadjoint operator $B := \sqrt{A} \in \mathcal{B}(H)$ such that $B^2 = A$, that is $\left(\sqrt{A}\right)^2 = A$. We call B the square root of A.

Let $A \in \mathcal{B}(H)$, then A^*A is selfadjoint and positive. Define the "operator absolute value" $|A| := \sqrt{A^*A}$. If $A = A^*$, then $|A| = \sqrt{A^2}$.

For a continuous real valued function φ we observe the following:

$$|\varphi(A)| \text{ (the functional absolute value) } = \int_{m-0}^{M} |\varphi(\lambda)| \, dE_\lambda =$$

$$\int_{m-0}^{M} \sqrt{(\varphi(\lambda))^2} dE_\lambda = \sqrt{(\varphi(A))^2} = |\varphi(A)| \text{ (operator absolute value)},$$

where A is a selfadjoint operator.

That is we have

$$|\varphi(A)| \text{ (functional absolute value) } = |\varphi(A)| \text{ (operator absolute value)}.$$

The next comes from [5], p. 3.

We say that a sequence $\{A_n\}_{n=1}^{\infty} \subset \mathcal{B}(H)$ converges uniformly to A (convergence in norm), iff

$$\lim_{n\to\infty} \|A_n - A\| = 0,$$

and w denote it as $\lim_{n\to\infty} A_n = A$.

We will be using Hölder's–McCarthy, 1967 ([15]), inequality: Let A be a selfadjoint positive operator on a Hilbert space H. Then

$$\langle A^r x, x \rangle \leq \langle Ax, x \rangle^r,$$

for all $0 < r < 1$ and $x \in H : \|x\| = 1$.

Let $A, B \in \mathcal{B}(H)$, then

$$\|AB\| \leq \|A\| \, \|B\|,$$

by Banach algebra property.

2.2 Main Results

Here we derive self adjoint operator-Korovkin type theorems via operator-Shisha–Mond type inequalities. This is a quantitative approach, studying the degree of operator-uniform approximation with rates of sequences of operator-positive linear operators in the operator order of $\mathcal{B}(H)$. We continue similarly with important polynomial operators. Our approach is direct based on Gelfand's isometry.

All the functions we are dealing here are real valued. We assume that $Sp(A) \subseteq [m, M]$.

Let $\{L_n\}_{n\in\mathbb{N}}$ be a sequence of positive linear operators from $C([m, M])$ into itself (i.e. if $f, g \in C([m, M])$ such that $f \geq g$, then $L_n(f) \geq L_n(g)$). It is interesting to study the convergence of $L_n \to I$ (unit operator, i.e. $I(f) = f, \forall f \in C([m, M])$). By property (i) we have that

$$\Phi(L_n f - f) = \Phi(L_n f) - \Phi(f) = (L_n f)(A) - f(A), \tag{2.1}$$

and

$$\Phi(L_n 1 \pm 1) = \Phi(L_n 1) \pm \Phi(1) = (L_n 1)(A) \pm 1_H, \tag{2.2}$$

the last comes by property (iv).

And by property (iii) we obtain

$$\|\Phi(L_n f - f)\| = \|(L_n f)(A) - f(A)\| = \|L_n f - f\|, \tag{2.3}$$

and

$$\|\Phi(L_n 1 \pm 1)\| = \|(L_n 1)(A) \pm 1_H\| = \|L_n(1) \pm 1\|. \tag{2.4}$$

We need

Theorem 2.1 (Shisha and Mond ([6]), 1968) *Let $\{L_n\}_{n\in\mathbb{N}}$ be a sequence of positive linear operators from $C([m, M])$ into itself. For $n = 1, 2, \ldots$, suppose $L_n(1)$ is bounded. Let $f \in C([m, M])$. Then for $n = 1, 2, \ldots$, we have*

$$\|L_n f - f\|_\infty \leq \|f\|_\infty \|L_n 1 - 1\|_\infty + \|L_n(1) + 1\|_\infty \omega_1(f, \mu_n), \tag{2.5}$$

where

$$\mu_n := \left\|L_n\left((t-x)^2\right)(x)\right\|_\infty^{\frac{1}{2}}, \tag{2.6}$$

with

$$\omega_1(f, \delta) := \sup_{\substack{x,y\in[m,M]\\ |x-y|\leq\delta}} |f(x) - f(y)|, \quad \delta > 0, \tag{2.7}$$

and $\|\cdot\|_\infty$ stands for the sup-norm over $[m, M]$.

In particular, if $L_n(1) = 1$, then (2.5) becomes

$$\|L_n(f) - f\|_\infty \leq 2\omega_1(f, \mu_n). \tag{2.8}$$

Note: (i) In foming μ_n^2, x is kept fixed, however t forms the functions t, t^2 on which L_n acts.

(ii) One can easily find, for $n = 1, 2, \ldots$,

$$\mu_n^2 \leq \left\| \left(L_n\left(t^2\right)\right)(x) - x^2 \right\|_\infty + 2c \left\|(L_n(t))(x) - x\right\|_\infty + c^2 \left\|(L_n(1))(x) - 1\right\|_\infty, \tag{2.9}$$

where

$$c := \max(|m|, |M|).$$

So, if the Korovkin's assumptions are fulfilled, i.e. if $L_n\left(id^2\right) \stackrel{u}{\to} id^2$, $L_n(id) \stackrel{u}{\to} id$ and $L_n(1) \stackrel{u}{\to} 1$, as $n \to \infty$, id is the identity map and u is the uniform convergence, then $\mu_n \to 0$, and then $\omega_1(f, \mu_n) \to 0$, as $n \to +\infty$, and we obtain from (2.5) that $\|L_n f - f\|_\infty \to 0$, i.e. $L_n(f) \stackrel{u}{\to} f$, as $n \to \infty$, $\forall f \in C([m, M])$.

We give

Theorem 2.2 *All as in Theorem 2.1. Then*

$$\|(L_n f)(A) - f(A)\| \leq$$

$$\|f(A)\| \, \|(L_n 1)(A) - 1_H\| + \|(L_n(1))(A) + 1_H\| \, \omega_1(f, \mu_n), \tag{2.10}$$

where

$$\mu_n := \left\| L_n\left((t - A)^2\right)(A) \right\|^{\frac{1}{2}}. \tag{2.11}$$

In particular, if $(L_n(1))(A) = 1_H$, then

$$\|(L_n(f))(A) - f(A)\| \leq 2\omega_1(f, \mu_n). \tag{2.12}$$

Furthermore it holds

$$\mu_n^2 \leq \left\| \left(L_n\left(t^2\right)\right)(A) - A^2 \right\| + 2c \left\|(L_n(t))(A) - A\right\| + c^2 \left\|(L_n(1))(A) - 1_H\right\|. \tag{2.13}$$

So, if $\left(L_n\left(t^2\right)\right)(A) \to A^2$, $(L_n(t))(A) \to A$, $(L_n(1))(A) \to 1_H$, uniformly, as $n \to \infty$, then by (2.10) and (2.13) we get $(L_n(f))(A) \to f(A)$, uniformly, as $n \to \infty$.

That is establishing the self adjoint operator Korovkin theorem with rates.

Next we follow [7], pp. 273–274.

Theorem 2.3 *Let $L_n : C([m, M]) \to C([m, M])$, $n \in \mathbb{N}$, be a sequence of positive linear operators, $f \in C([m, M])$, $g \in C([m, M])$ and it is an $(1-1)$ function.*

Assume $\{L_n(1)\}_{n\in\mathbb{N}}$ *is uniformly bounded. Then*

$$\|L_n(f) - f\| \leq \|f\| \|L_n(1) - 1\| + (1 + \|L_n(1)\|)\, \omega_g(f, \rho_n), \tag{2.14}$$

where

$$\omega_g(f, h) := \sup_{x,y} \{|f(x) - f(y)| : |g(x) - g(y)| \leq h\}, \tag{2.15}$$

$h > 0$, *with*

$$\rho_n := \left(\left\|L_n\left((g - g(y))^2\right)(y)\right\|\right)^{\frac{1}{2}}. \tag{2.16}$$

Here $\|\cdot\|$ *stands for the supremum norm. If* $L_n(1) = 1$, *then (2.14) simplifies to*

$$\|L_n(f) - f\| \leq 2\omega_g(f, \rho_n). \tag{2.17}$$

We also have that

$$\rho_n^2 \leq \left\|L_n\left(g^2\right) - g^2\right\| + 2\|g\| \|L_n(g) - g\| + \|g\|^2 \|L_n(1) - 1\|. \tag{2.18}$$

If $L_n(1) \overset{u}{\to} 1$, $L_n(g) \overset{u}{\to} g$, $L_n\left(g^2\right) \overset{u}{\to} g^2$, *then* $\omega_g(f, \rho_n) \to 0$, *and then* $L_n(f) \overset{u}{\to} f$, *as* $n \to +\infty, \forall f \in C([m, M])$, *where* u *stands for uniform convergence, so we get a generalization of Korovkin theorem quantitatively, and clearly by* $L_n(1) \overset{u}{\to} 1$, *we get* $\|L_n(1)\| \leq K$, $\forall\, n \in \mathbb{N}$, *where* $K > 0$.

We present

Theorem 2.4 *All as in Theorem 2.3. Then*

$$\|(L_n(f))(A) - f(A)\| \leq$$

$$\|f(A)\| \|(L_n(1))(A) - 1_H\| + (1 + \|(L_n(1))(A)\|)\, \omega_g(f, \rho_n), \tag{2.19}$$

with

$$\rho_n := \left(\left\|L_n\left((g - g(A))^2\right)(A)\right\|\right)^{\frac{1}{2}}. \tag{2.20}$$

If $(L_n(1))(A) = 1_H$, *then*

$$\|(L_n(f))(A) - f(A)\| \leq 2\omega_g(f, \rho_n). \tag{2.21}$$

It holds

$$\rho_n^2 \leq \left\|\left(L_n\left(g^2\right)\right)(A) - g^2(A)\right\| +$$

$$2\|g(A)\| \|(L_n(g))(A) - A\| + \|g(A)\|^2 \|(L_n(1))(A) - 1_H\|. \tag{2.22}$$

If $(L_n(1))(A) \to 1_H$, $(L_n(g))(A) \to A$, $\left(L_n\left(g^2\right)\right)(A) \to g^2(A)$, *uniformly, as* $n \to +\infty$, *then* $(L_n(f))(A) \to f(A)$, *uniformly, as* $n \to +\infty$.

We make

Remark 2.5 Next we consider the general Bernstein positive linear polynomial operators from $C([m, M])$ into itself, for $f \in C([m, M])$ we define

$$(B_N f)(s) = \sum_{i=0}^{N} \binom{N}{i} f\left(m + i\left(\frac{M-m}{N}\right)\right)\left(\frac{s-m}{M-m}\right)^i \left(\frac{M-s}{M-m}\right)^{N-i}, \tag{2.23}$$

$\forall\, s \in [m, M]$, see [8], p. 80.

Then by [8], p. 81, we get that

$$\|B_N f - f\|_\infty \leq \frac{5}{4}\omega_1\left(f, \frac{M-m}{\sqrt{N}}\right), \tag{2.24}$$

$\forall\, N \in \mathbb{N}$, i.e. $B_N f \overset{u}{\to} f$, as $N \to +\infty$, $\forall\, f \in C([m, M])$, the convergence is given with rates.

We clearly have that

$$\|(B_N f)(A) - f(A)\| \leq \frac{5}{4}\omega_1\left(f, \frac{M-m}{\sqrt{N}}\right), \tag{2.25}$$

$\forall\, N \in \mathbb{N}$, i.e. $(B_N f)(A) \to f(A)$, uniformly, as $N \to +\infty$.

We need

Notation 2.6 *Let* $x \in [m, M]$. *Denote*

$$c(x) := \max(x - m, M - x) = \frac{1}{2}\left[M - m + |M + m - 2x|\right] > 0. \tag{2.26}$$

Let $h > 0$ *be fixed,* $n \in \mathbb{N}$. *Define (see [9], p. 210)*

$$\Phi_{*n}(x) := \left(\frac{|x|^{n+1}}{(n+1)!h} + \frac{|x|^n}{2n!} + \frac{h\,|x|^{n-1}}{8(n-1)!}\right). \tag{2.27}$$

We need

Theorem 2.7 ([9], p. 219) *Let* $\{L_N\}_{N \in \mathbb{N}}$ *be a sequence of positive linear operators from* $C([m, M])$ *into itself,* $x \in [m, M]$, $f \in C^n([m, M])$.

Here $c(x)$, $\Phi_{*n}(x)$ as in Notation 2.6. Assume that $\omega_1\left(f^{(n)}, h\right) \leq w$, where w, h are fixed positive numbers, $0 < h < M - m$. Then

$$|(L_N(f))(x) - f(x)| \leq |f(x)|\,|(L_N(1))(x) - 1| +$$

$$\sum_{k=1}^{n} \frac{\left|f^{(k)}(x)\right|}{k!} \left|\left(L_N\left((t-x)^k\right)\right)(x)\right| + \frac{w\Phi_{*n}(c(x))}{(c(x))^n}\left(L_N\left(|t-x|^n\right)\right)(x). \quad (2.28)$$

Inequality (2.28) is sharp, for details see [9], p. 220.

Clearly all functions involved in (2.28) are continuous, see also [10], i.e. both sides of (2.28) are continuous functions.

Using properties (P) and (ii) and (2.28) we derive

Theorem 2.8 *All as in Theorem 2.7. Then*

$$|(L_N(f))(A) - f(A)| \leq |f(A)|\,|(L_N(1))(A) - 1_H| +$$

$$\sum_{k=1}^{n} \frac{\left|f^{(k)}(A)\right|}{k!} \left|\left(L_N\left((t-A)^k\right)\right)(A)\right| + \frac{w\Phi_{*n}(c(A))}{(c(A))^n}\left(L_N\left(|t-A|^n\right)\right)(A). \quad (2.29)$$

Remark 2.9 Inequality (2.29) implies

$$\|(L_N(f))(A) - f(A)\| \leq \|f(A)\|\,\|(L_N(1))(A) - 1_H\| +$$

$$\sum_{k=1}^{n} \frac{\left\|f^{(k)}(A)\right\|}{k!} \left\|\left(L_N\left((t-A)^k\right)\right)(A)\right\| + w\left\|\frac{\Phi_{*n}(c(A))}{(c(A))^n}\right\|\left\|\left(L_N\left(|t-A|^n\right)\right)(A)\right\|. \quad (2.30)$$

Remark 2.10 (to Theorem 2.8 *and* (2.30)*)* Assume further

$$\|L_N(1)\|_\infty \leq \mu, \ \forall\, N \in \mathbb{N};\ \mu > 0. \quad (2.31)$$

By Riesz representation theorem, for each $s \in [m, M]$, there exists a positive finite measure μ_s on $[m, M]$ such that

$$(L_N(f))(s) = \int_{[m,M]} f(t)\, d\mu_{sN}(t),\ \forall\, f \in C([m, M]). \quad (2.32)$$

Therefore $(k = 1, \ldots, n-1)$

$$\left|\left(L_N(\cdot - s)^k\right)(s)\right| = \left|\int_{[m,M]} (\lambda - s)^k\, d\mu_{sN}(\lambda)\right| \leq \int_{[m,M]} |\lambda - s|^k\, d\mu_{sN}(\lambda)$$

(by Hölder's inequality)

$$\leq \left(\int_{[m,M]} 1 d\mu_{sN}(\lambda) \right)^{\frac{n-k}{n}} \left(\int_{[m,M]} |\lambda - s|^n d\mu_{sN}(\lambda) \right)^{\frac{k}{n}}$$

$$= ((L_N(1))(s))^{\frac{n-k}{n}} \left((L_N(|\cdot - s|^n))(s) \right)^{\frac{k}{n}} \leq \mu^{\frac{n-k}{n}} \left((L_N(|\cdot - s|^n))(s) \right)^{\frac{k}{n}}. \quad (2.33)$$

That is

$$\left| \left(L_N(\cdot - s)^k \right)(s) \right| \leq \mu^{\frac{n-k}{n}} \left((L_N(|\cdot - s|^n))(s) \right)^{\frac{k}{n}}, \quad (2.34)$$

$k = 1, \ldots, n-1$.

Of course it holds

$$\left| \left(L_N(\cdot - s)^n \right)(s) \right| \leq \left(L_N |\cdot - s|^n \right)(s). \quad (2.35)$$

By property (P) we obtain

$$\left| \left(L_N(\cdot - A)^k \right)(A) \right| \leq \mu^{\frac{n-k}{n}} \left((L_N(|\cdot - A|^n))(A) \right)^{\frac{k}{n}}, \quad (2.36)$$

for $k = 1, \ldots, n-1$, and

$$\left| \left(L_N(\cdot - A)^n \right)(A) \right| \leq \left(L_N |\cdot - A|^n \right)(A). \quad (2.37)$$

Therefore

$$\left\| \left(L_N(\cdot - A)^k \right)(A) \right\| \leq \mu^{\frac{n-k}{n}} \left\| \left((L_N(|\cdot - A|^n))(A) \right)^{\frac{k}{n}} \right\| \quad (2.38)$$

$$= \mu^{\frac{n-k}{k}} \sup_{x \in H : \|x\| = 1} \left\langle \left((L_N(|\cdot - A|^n))(A) \right)^{\frac{k}{n}} x, x \right\rangle$$

(by Hölder's–Mc Carthy inequality)

$$\leq \mu^{\frac{n-k}{k}} \sup_{x \in H : \|x\| = 1} \left\langle \left((L_N(|\cdot - A|^n))(A) \right) x, x \right\rangle^{\frac{k}{n}}$$

$$= \mu^{\frac{n-k}{k}} \left(\sup_{x \in H : \|x\| = 1} \left\langle \left((L_N(|\cdot - A|^n))(A) \right) x, x \right\rangle \right)^{\frac{k}{n}}$$

$$= \mu^{\frac{n-k}{k}} \left\| \left(L_N(|\cdot - A|^n) \right)(A) \right\|^{\frac{k}{n}}. \quad (2.39)$$

Therefore it holds

$$\left\| \left(L_N(t - A)^k \right)(A) \right\| \leq \mu^{\frac{n-k}{k}} \left\| \left(L_N(|t - A|^n) \right)(A) \right\|^{\frac{k}{n}}, \quad (2.40)$$

$k = 1, \ldots, n-1$, and of course

$$\left\| \left(L_N (t-A)^n\right)(A) \right\| \leq \left\| \left(L_N \left(|t-A|^n\right)\right)(A) \right\|. \tag{2.41}$$

Based on (2.40) and (2.41) and by assuming that $(L_n(1))(A) \to 1_H$ and $(L_N (|t-A|^n))(A) \to 0_H$, uniformly, as $N \to +\infty$, we obtain by (2.30) that $(L_N(f))(A) \to f(A)$, uniformly as $N \to +\infty$.

We mention

Theorem 2.11 ([9], p. 230) *For any $f \in C^1([0,1])$ consider the Bernstein polynomials*

$$(\beta_n(f))(t) := \sum_{k=0}^{n} f\left(\frac{k}{n}\right)\binom{n}{k} t^k (1-t)^{n-k}, \ t \in [0,1].$$

Then

$$\|(\beta_n f) - f\|_\infty \leq \frac{0.78125}{\sqrt{n}} \omega_1 \left(f', \frac{1}{4\sqrt{n}}\right). \tag{2.42}$$

We make

Remark 2.12 The map

$$[m, M] \ni s = \varphi(t) = (M-m)t + m, \ t \in [0,1], \tag{2.43}$$

maps $(1-1)$ and onto, $[0,1]$ onto $[m, M]$.

Let $f \in C^1([m, M])$, then

$$f(s) = f(\varphi(t)) = f((M-m)t + m), \tag{2.44}$$

and

$$\frac{df(s)}{dt} = (f(\varphi(t)))' = f'(\varphi(t))(M-m) = f'(s)(M-m). \tag{2.45}$$

By (2.42) we get that

$$\|\beta_n(f((M-m))t+m) - f((M-m)t+m)\|_{\infty,[0,1]} \leq$$

$$\frac{0.78125}{\sqrt{n}} \omega_1 \left(f'(s)(M-m), \frac{1}{4\sqrt{n}}\right) = \frac{0.78125}{\sqrt{n}} (M-m)\, \omega_1 \left(f'(s), \frac{1}{4\sqrt{n}}\right). \tag{2.46}$$

However we have

$$\omega_1\left(f'(s), \frac{1}{4\sqrt{n}}\right) = \omega_1\left(f'((M-m)t+m), \frac{1}{4\sqrt{n}}\right) = \tag{2.47}$$

$$\sup_{\substack{t_1,t_2\in[0,1] \\ |t_1-t_2|\le\frac{1}{4\sqrt{n}}}} \left|f'((M-m)t_1+m) - f'((M-m)t_2+m)\right| =$$

$$\sup_{\substack{s_1,s_2\in[m,M] \\ |s_1-s_2|\le\frac{M-m}{4\sqrt{n}}}} \left|f'(s_1) - f'(s_2)\right| = \omega_1\left(f', \frac{M-m}{4\sqrt{n}}\right), \tag{2.48}$$

above notice that

$$|s_1 - s_2| = |((M-m)t_1+m) - ((M-m)t_2+m)| =$$

$$(M-m)|t_1-t_2| \le \frac{M-m}{4\sqrt{n}}. \tag{2.49}$$

So we have proved that

$$\omega_1\left(f'(s), \frac{1}{4\sqrt{n}}\right) = \omega_1\left(f', \frac{M-m}{4\sqrt{n}}\right). \tag{2.50}$$

Finally, we observe that

$$(\beta_n(f((M-m)t+m))(t)) =$$

$$\sum_{k=0}^{n}\left(f\left((M-m)\frac{k}{n}+m\right)\right)\binom{n}{k}t^k(1-t)^{n-k} =$$

$$\sum_{k=0}^{n}\left(f\left((M-m)\frac{k}{n}+m\right)\right)\binom{n}{k}\left(\frac{s-m}{M-m}\right)^k\left(\frac{M-s}{M-m}\right)^{n-k} =: (B_n(f))(s), \tag{2.51}$$

$s \in [m, M]$.

The operators $(B_n(f))(s)$ are the general Bernstein polynomials.

From (2.46) and (2.50), we derive that

$$\|(B_n f)(s) - f(s)\|_{\infty,[m,M]} \le \frac{0.78125}{\sqrt{n}}(M-m)\,\omega_1\left(f', \frac{M-m}{4\sqrt{n}}\right). \tag{2.52}$$

Based on the above and the property (iii), we can give

Theorem 2.13 *Let $f' \in [m, M]$. Then*

$$\|(B_n f)(A) - f(A)\| \leq \frac{0.78125\,(M-m)}{\sqrt{n}} \omega_1 \left(f', \frac{M-m}{4\sqrt{n}} \right). \tag{2.53}$$

I.e. $(B_n f)(A) \to A$, uniformly, with rates as $n \to +\infty$.

We make

Remark 2.14 Let $f \in C([m, M])$, then the function $f((M-m)t+m)$ is a continuous function in $t \in [0, 1]$.

Let $r \in \mathbb{N}$, we evaluate the modulus of smoothness ($\delta > 0$)

$$\omega_r (f((M-m)t+m), \delta) =$$

$$\sup_{0 \leq h \leq \delta} \left\| \sum_{k=0}^{r} \binom{r}{k} (-1)^{r-k} f((M-m)(t+kh)+m) \right\|_{\infty,[0,1-rh]} =$$

$$\sup_{0 \leq h^* \leq \delta(M-m)} \left\| \sum_{k=0}^{r} \binom{r}{k} (-1)^{r-k} f\left(s+kh^*\right) \right\|_{s,\infty,[m,M-rh^*]}$$

$(h^* = (M-m)h)$

$$= \omega_r (f, (M-m)\delta), \tag{2.54}$$

proving that

$$\omega_r (f((M-m)t+m), \delta) = \omega_r (f, (M-m)\delta), \tag{2.55}$$

for any $r \in \mathbb{N}$, and $\delta > 0$.

We need

Theorem 2.15 ([11], p.97) *For $f \in C([0, 1])$, $n \in \mathbb{N}$, we have*

$$\|\beta_n(f) - f\| \leq \omega_2 \left(f, \frac{1}{\sqrt{n}} \right), \tag{2.56}$$

a sharp inequality.

We get

Theorem 2.16 *Let $f \in C([m, M])$, $n \in \mathbb{N}$. Then*

$$\|(B_n(f))(A) - f(A)\| = \|B_n(f) - f\|_\infty \leq \omega_2 \left(f, \frac{M-m}{\sqrt{n}} \right). \tag{2.57}$$

We need

Definition 2.17 *([11], p. 151)* Let $f \in C([0,1])$, $n \in \mathbb{N}$. We define the Durrmeyer type operators (the genuine Bernstein-Durrmeyer operators)

$$\left(M_n^{-1,-1}(f)\right)(x) = f(0)(1-x)^n + f(1)x^n +$$

$$(n-1)\sum_{k=1}^{n-1} p_{n,k}(x)\int_0^1 f(t)\,p_{n-2,k-1}(t)\,dt, \tag{2.58}$$

where

$$p_{n,k}(x) = \binom{n}{k} x^k (1-x)^{n-k}, \ n \in \mathbb{N}, \ x \in [0,1].$$

We will use

Theorem 2.18 *([11], p. 155) For $f \in C([0,1])$, $n \in \mathbb{N}$, we have*

$$\left\|M_n^{-1,-1}(f) - f\right\|_\infty \le \frac{5}{4}\omega_2\left(f, \frac{1}{\sqrt{n+1}}\right). \tag{2.59}$$

We make

Remark 2.19 Let $f \in C([m,M])$, then $f((M-m)t+m) \in C([0,1])$. Hence ($s \in [m,M]$, $t \in [0,1]$)

$$\left(\overline{M}_n^{-1,-1} f\right)(s) := M_n^{-1,-1}\left(f((M-m)t+m)\right)(t) \overset{(2.58)}{=}$$

$$f(m)\left(\frac{M-s}{M-m}\right)^n + f(M)\left(\frac{s-m}{M-m}\right)^n +$$

$$(n-1)\sum_{k=1}^{n-1}\binom{n}{k}\left(\frac{s-m}{M-m}\right)^k\left(\frac{M-s}{M-m}\right)^{n-k}\binom{n-2}{k-1}\int_m^M f(\overline{s})\,\cdot$$

$$\left(\frac{\overline{s}-m}{M-m}\right)^{k-1}\left(\frac{M-\overline{s}}{M-m}\right)^{n-k-1} d\overline{s}. \tag{2.60}$$

We give

Theorem 2.20 *Let $f \in C([m,M])$, $n \in \mathbb{N}$. Then*

$$\left\|\left(\overline{M}_n^{-1,-1} f\right)(A) - f(A)\right\| = \left\|\overline{M}_n^{-1,-1} f - f\right\|_\infty \le \frac{5}{4}\omega_2\left(f, \frac{M-m}{\sqrt{n+1}}\right). \tag{2.61}$$

We need

Definition 2.21 *([12])* For $f \in C([0,1])$, $w \in \mathbb{N}$, and $0 \leq \beta \leq \gamma$, we define the Stancu-type positive linear operators

$$\left(L_{w_0}^{\langle 0\beta\gamma\rangle} f\right)(x) = \sum_{k=0}^{w} f\left(\frac{k+\beta}{w+\gamma}\right) p_{w,k}(x), \; x \in [0,1], \tag{2.62}$$

$$p_{w,k}(x) = \binom{w}{k} x^k (1-x)^{w-k}.$$

We need

Theorem 2.22 ([7], p.516 and [12]) *For $\mathbb{N} \ni w > \lceil \gamma^2 \rceil$ ($\lceil \cdot \rceil$ is the ceiling), $f \in C([0,1])$ we have:*

$$\left\| L_{w_0}^{\langle 0\beta\gamma\rangle} f - f \right\|_\infty \leq \left[3 + \frac{\left(w^3 + 4w\beta^2 \left(w - \gamma^2\right)\right)}{4\left(w-\gamma^2\right)(w+\gamma)^2} \right] \omega_2\left(f, \frac{1}{\sqrt{w}}\right)$$

$$+ \frac{2(\beta+\gamma)\sqrt{w}}{(w+\gamma)} \omega_1\left(f, \frac{1}{\sqrt{w}}\right). \tag{2.63}$$

We make

Remark 2.23 Let $f \in C([m,M])$, then $f((M-m)t+m) \in C([0,1])$. Hence $(s \in [m,M], t \in [0,1])$

$$\left(\overline{L}_{w_0}^{\langle 0\beta\gamma\rangle} f\right)(s) := L_{w_0}^{\langle 0\beta\gamma\rangle} \left(f((M-m)t+m)\right)(t) \overset{(2.62)}{=}$$

$$\sum_{k=0}^{w} f\left((M-m)\left(\frac{k+\beta}{w+\gamma}\right)+m\right) \binom{w}{k} \left(\frac{s-m}{M-m}\right)^k \left(\frac{M-s}{M-m}\right)^{w-k}. \tag{2.64}$$

We give

Theorem 2.24 *Let $f \in C([m,M])$, $w \in \mathbb{N}$, $0 \leq \beta \leq \gamma$. We take $w > \lceil \gamma^2 \rceil$. Then*

$$\left\| \left(\overline{L}_{w_0}^{\langle 0\beta\gamma\rangle} f\right)(A) - f(A) \right\| = \left\| \left(\overline{L}_{w_0}^{\langle 0\beta\gamma\rangle} f\right) - f \right\|_\infty \leq$$

$$\left[3 + \frac{\left(w^3 + 4w\beta^2 \left(w-\gamma^2\right)\right)}{4\left(w-\gamma^2\right)(w+\gamma)^2} \right] \omega_2\left(f, \frac{M-m}{\sqrt{w}}\right) + \frac{2(\beta+\gamma)\sqrt{w}}{(w+\gamma)} \omega_1\left(f, \frac{M-m}{\sqrt{w}}\right). \tag{2.65}$$

We make

Remark 2.25 Next we assume that the spectrum of A is $[0,1]$. For example, it could be $Af = xf(x)$ on $L^2([0,1])$ which is a self adjoint operator and it has spectrum $[0,1]$.

We need

Definition 2.26 *([13])* Let $f \in C([0,1])$, we define the special Stancu operator

$$S_n(f,x) = \frac{2(n!)}{(2n)!} \sum_{k=0}^{n} f\left(\frac{k}{n}\right) \binom{n}{k} (nx)_k (n-nx)_{n-k}, \tag{2.66}$$

where $(a)_0 = 1$, $(a)_b = \sum_{k=0}^{b-1} (a-k)$, $a \in \mathbb{R}$, $b \in \mathbb{N}$, $n \in \mathbb{N}$, $x \in [0,1]$.

Theorem 2.27 ([14], p. 75) *Let $f \in C([0,1])$, $n \in \mathbb{N}$. Then*

$$\left| \left(\left(S_n - M_n^{-1,-1} \right)(f) \right)(x) \right| \le c_1 \omega_4 \left(f, \sqrt[4]{\frac{3x(1-x)}{n(n+1)}} \right), \tag{2.67}$$

$\forall\, x \in [0,1]$, *where $c_1 > 0$ is an absolute constant independent of n, f and x.*

We obtain

Theorem 2.28 *Let $f \in C([0,1])$, $n \in \mathbb{N}$. Then*

$$\left\| \left(S_n - M_n^{-1,-1} \right)(A) \right\| = \left\| S_n - M_n^{-1,-1} \right\|_\infty \le c_1 \omega_4 \left(f, \sqrt[4]{\frac{3}{4n(n+1)}} \right). \tag{2.68}$$

References

1. G. Anastassiou, *Self Adjoint Operator Korovkin Type and Polynomial Direct Approximations with Rates*, (2016, submitted)
2. T. Furuta, J. Mićić Hot, J. Pečaric, Y. Seo, *Mond-Pečaric Method in Operator Inequalities. Inequalities for Bounded Selfadjoint Operators on a Hilbert Space*, Element, Zagreb (2005)
3. S. Dragomir, *Operator Inequalities of Ostrowski and Trapezoidal Type* (Springer, New York, 2012)
4. G. Helmberg, *Introduction to Spectral Thery in Hilbert Space* (Wiley, New York, 1969)
5. S.S. Dragomir, Inequalities for functions of selfadjoint operators on Hilbert Spaces (2011), ajmaa.org/RGMIA/monographs/InFuncOp.pdf
6. O. Shisha, B. Mond, The degree of convergence of sequences of linear positive operators. Nat. Acad. Sci. U.S. **60**, 1196–1200 (1968)
7. G.A. Anastassiou, *Intelligent Mathematics: Computational Analysis* (Springer, Heidelberg, 2011)
8. L. Shumaker, *Spline Functions Basic Theory* (Wiley-Interscience, New York, 1981)
9. G.A. Anastassiou, *Moments in Probability and Approximation Theory*, vol. 287, Longman Scientific & Technical, Pitman Research Notes in Mathematics Series (Wiley, New York, 1993)
10. G.A. Anastassiou, Self adjoint operator Korovkin type quantitative approximations, Acta Mathematica Universitatis Comenianae (2016, accepted)
11. R. Paltanea, *Aproximation Theory Using Positive Linear Operators* (Birkhauser, Boston, 2004)

12. H.H. Gonska, J. Meier, *Quantitative Theorems on Approximation by Bernstein-Stancu Operators*, Estratto da Calcolo 21 (fasc. IV), 317–335, 1984
13. D.D. Stancu, Approximation of functions by a new class of linear polynomial operators. Rev. Roumaine Math. Pures Appl. **13**, 1173–1194 (1968)
14. H. Gonska, P. Pitul, I. Rasa, On Peano's form of the Taylor remainder, Voronovskaja's theorem and the commutator of positive linear operators, in *Proccedings of the International Conference on Numerical Analysis and Approximation theory*, ed. by O. Agratini, P. Blaga, Babes-Bolyai University, Cluj-Napoca, Romania, 5–8 July 2006, pp. 55–80
15. C.A. McCarthy, c_p. Israel J. Math. **5**, 249–271 (1967)

Chapter 3
Quantitative Self Adjoint Operator Other Direct Approximations

Here we give a series of self adjoint operator positive linear operators general results. Then we present specific similar results related to neural networks. This is a quantitative treatment to determine the degree of self adjoint operator uniform approximation with rates, of sequences of self adjoint positive linear operators in general, and in particular of self adjoint specific neural network operators. The approach is direct relying on Gelfand's isometry. It follows [4].

3.1 Background

Let A be a selfadjoint linear operator on a complex Hilbert space $(H; \langle \cdot, \cdot \rangle)$. The Gelfand map establishes a $*-$isometrically isomorphism Φ between the set $C(Sp(A))$ of all continuous functions defined on the spectrum of A, denoted $Sp(A)$, and the C^*-algebra $C^*(A)$ generated by A and the identity operator 1_H on H as follows (see e.g. [8, p. 3]):

For any $f, g \in C(Sp(A))$ and any $\alpha, \beta \in \mathbb{C}$ we have

(i) $\Phi(\alpha f + \beta g) = \alpha \Phi(f) + \beta \Phi(g)$;

(ii) $\Phi(fg) = \Phi(f)\Phi(g)$ (the operation composition is on the right) and $\Phi(\overline{f}) = (\Phi(f))^*$;

(iii) $\|\Phi(f)\| = \|f\| := \sup_{t \in Sp(A)} |f(t)|$;

(iv) $\Phi(f_0) = 1_H$ and $\Phi(f_1) = A$, where $f_0(t) = 1$ and $f_1(t) = t$, for $t \in Sp(A)$.

With this notation we define

$$f(A) := \Phi(f), \text{ for all } f \in C(Sp(A)),$$

and we call it the continuous functional calculus for a selfadjoint operator A.

G.A. Anastassiou, *Intelligent Comparisons II: Operator Inequalities and Approximations*, Studies in Computational Intelligence 699,
DOI 10.1007/978-3-319-51475-8_3

If A is a selfadjoint operator and f is a real valued continuous function on $Sp(A)$ then $f(t) \geq 0$ for any $t \in Sp(A)$ implies that $f(A) \geq 0$, i.e. $f(A)$ is a positive operator on H. Moreover, if both f and g are real valued continuous functions on $Sp(A)$ then the following important property holds:

(P) $f(t) \geq g(t)$ for any $t \in Sp(A)$, implies that $f(A) \geq g(A)$ in the operator order of $B(H)$.

Equivalently, we use (see [7], pp. 7–8):

Let U be a selfadjoint operator on the complex Hilbert space $(H, \langle \cdot, \cdot \rangle)$ with the spectrum $Sp(U)$ included in the interval $[m, M]$ for some real numbers $m < M$ and $\{E_\lambda\}_\lambda$ be its spectral family.

Then for any continuous function $f : [a, b] \to \mathbb{C}$, where $[m, M] \subset (a, b)$, it is well known that we have the following spectral representation in terms of the Riemann—Stieljes integral:

$$\langle f(U)x, y \rangle = \int_{m-0}^{M} f(\lambda)\, d(\langle E_\lambda x, y \rangle),$$

for any $x, y \in H$. The function $g_{x,y}(\lambda) := \langle E_\lambda x, y \rangle$ is of bounded variation on the interval $[m, M]$, and

$$g_{x,y}(m-0) = 0 \text{ and } g_{x,y}(M) = \langle x, y \rangle,$$

for any $x, y \in H$. Furthermore, it is known that $g_x(\lambda) := \langle E_\lambda x, x \rangle$ is increasing and right continuous on $[m, M]$.

In this chapter we will be using a lot the formula

$$\langle f(U)x, x \rangle = \int_{m-0}^{M} f(\lambda)\, d(\langle E_\lambda x, x \rangle), \forall\, x \in H.$$

As a symbol we can write

$$f(U) = \int_{m-0}^{M} f(\lambda)\, dE_\lambda.$$

Above, $m = \min\{\lambda | \lambda \in Sp(U)\} := \min Sp(U)$, $M = \max\{\lambda | \lambda \in Sp(U)\} := \max Sp(U)$. The projections $\{E_\lambda\}_{\lambda \in \mathbb{R}}$, are called the spectral family of A, with the properties:

(a) $E_\lambda \leq E_{\lambda'}$ for $\lambda \leq \lambda'$;
(b) $E_{m-0} = 0_H$ (zero operator), $E_M = 1_H$ (identity operator) and $E_{\lambda+0} = E_\lambda$ for all $\lambda \in \mathbb{R}$.

Furthermore

$$E_\lambda := \varphi_\lambda(U), \ \forall\, \lambda \in \mathbb{R},$$

is a projection which reduces U, with

$$\varphi_\lambda(s) := \begin{cases} 1, \text{ for } -\infty < s \leq \lambda, \\ 0, \text{ for } \lambda < s < +\infty. \end{cases}$$

The spectral family $\{E_\lambda\}_{\lambda \in \mathbb{R}}$ determines uniquely the self-adjoint operator U and vice versa.

For more on the topic see [9], pp. 256–266, and for more details see there pp. 157–266. See also [6].

Some more basics are given (we follow [7], pp. 1–5):

Let $(H; \langle \cdot, \cdot \rangle)$ be a Hilbert space over $\mathbb{C}$. A bounded linear operator A defined on H is selfjoint, i.e., $A = A^*$, iff $\langle Ax, x \rangle \in \mathbb{R}$, $\forall\, x \in H$, and if A is selfadjoint, then

$$\|A\| = \sup_{x \in H: \|x\|=1} |\langle Ax, x \rangle| .$$

Let A, B be selfadjoint operators on H. Then $A \leq B$ iff $\langle Ax, x \rangle \leq \langle Bx, x \rangle, \forall\, x \in H$.

In particular, A is called positive if $A \geq 0$.

Denote by

$$\mathcal{P} := \left\{ \varphi(s) := \sum_{k=0}^{n} \alpha_k s^k | n \geq 0, \alpha_k \in \mathbb{C}, 0 \leq k \leq n \right\}.$$

If $A \in \mathcal{B}(H)$ (the Banach algebra of all bounded linear operators defined on H, i.e. from H into itself) is selfadjoint, and $\varphi(s) \in \mathcal{P}$ has real coefficients, then $\varphi(A)$ is selfadjoint, and

$$\|\varphi(A)\| = \max\{|\varphi(\lambda)|, \lambda \in Sp(A)\}.$$

If φ is any function defined on $\mathbb{R}$ we define

$$\|\varphi\|_A := \sup\{|\varphi(\lambda)|, \lambda \in Sp(A)\}.$$

If A is selfadjoint operator on Hilbert space H and φ is continuous and given that $\varphi(A)$ is selfadjoint, then $\|\varphi(A)\| = \|\varphi\|_A$. And if φ is a continuous real valued function so it is $|\varphi|$, then $\varphi(A)$ and $|\varphi|(A) = |\varphi(A)|$ are selfadjoint operators (by [7], p. 4, Theorem 7).

Hence it holds

$$\| |\varphi(A)| \| = \| |\varphi| \|_A = \sup\{ ||\varphi(\lambda)||, \lambda \in Sp(A)\}$$

$$= \sup\{|\varphi(\lambda)|, \lambda \in Sp(A)\} = \|\varphi\|_A = \|\varphi(A)\|,$$

that is

$$\| |\varphi(A)| \| = \|\varphi(A)\|.$$

For a selfadjoint operator $A \in \mathcal{B}(H)$ which is positive, there exists a unique positive selfadjoint operator $B := \sqrt{A} \in \mathcal{B}(H)$ such that $B^2 = A$, that is $(\sqrt{A})^2 = A$. We call B the square root of A.

Let $A \in \mathcal{B}(H)$, then A^*A is selfadjoint and positive. Define the "operator absolute value" $|A| := \sqrt{A^*A}$. If $A = A^*$, then $|A| = \sqrt{A^2}$.

For a continuous real valued function φ we observe the following:

$$|\varphi(A)| \text{ (the functional absolute value)} = \int_{m-0}^{M} |\varphi(\lambda)| \, dE_\lambda =$$

$$\int_{m-0}^{M} \sqrt{(\varphi(\lambda))^2} dE_\lambda = \sqrt{(\varphi(A))^2} = |\varphi(A)| \text{ (operator absolute value)},$$

where A is a selfadjoint operator.

That is we have

$$|\varphi(A)| \text{ (functional absolute value)} = |\varphi(A)| \text{ (operator absolute value)}.$$

The next comes from [6], p. 3:

We say that a sequence $\{A_n\}_{n=1}^{\infty} \subset \mathcal{B}(H)$ converges uniformly to A (convergence in norm), iff

$$\lim_{n\to\infty} \|A_n - A\| = 0,$$

and w denote it as $\lim_{n\to\infty} A_n = A$.

We will be using Hölder's—McCarthy, 1967 ([10]), inequality: Let A be a selfadjoint positive operator on a Hilbert space H. Then

$$\langle A^r x, x\rangle \le \langle Ax, x\rangle^r,$$

for all $0 < r < 1$ and $x \in H : \|x\| = 1$.

Let $A, B \in \mathcal{B}(H)$, then

$$\|AB\| \le \|A\| \, \|B\|,$$

by Banach algebra property.

3.2 Main Results

Our approach is direct based on Gelfand's isometry.

All the functions we are dealing here are real valued. We assume that $Sp(A) \subseteq [m, M]$.

Let $\{L_n\}_{n\in\mathbb{N}}$ be a sequence of positive linear operators from $C([m, M])$ into itself (i.e. if $f, g \in C([m, M])$ such that $f \geq g$, then $L_n(f) \geq L_n(g)$). It is interesting to study the convergence of $L_n \to I$ (unit operator, i.e. $I(f) = f, \forall f \in C([m, M])$). By property (i) we have that

$$\Phi(L_n f - f) = \Phi(L_n f) - \Phi(f) = (L_n f)(A) - f(A), \tag{3.1}$$

and

$$\Phi(L_n 1 \pm 1) = \Phi(L_n 1) \pm \Phi(1) = (L_n 1)(A) \pm 1_H, \tag{3.2}$$

the last comes by property (iv).

And by property (iii) we obtain

$$\|\Phi(L_n f - f)\| = \|(L_n f)(A) - f(A)\| = \|L_n f - f\|, \tag{3.3}$$

and

$$\|\Phi(L_n 1 \pm 1)\| = \|(L_n 1)(A) \pm 1_H\| = \|L_n(1) \pm 1\|. \tag{3.4}$$

We need the first modulus of continuity

$$\omega_1(f, \delta) := \sup_{\substack{x,y\in[m,M] \\ |x-y|\leq\delta}} |f(x) - f(y)|, \quad \delta > 0, \tag{3.5}$$

here $\|\cdot\|_\infty$ stands for the sup-norm over $[m, M]$.

We need and mention

Theorem 3.1 ([1], p. 419) *Consider the positive linear operator*

$$L : C([m, M]) \to C([m, M]).$$

Define $(n \in \mathbb{N})$

$$D_n := \left\|\left(L\left(|t - \cdot|^n\right)\right)(\cdot)\right\|_\infty^{\frac{1}{n}}. \tag{3.6}$$

Let $f \in C^n([m, M])$. *Then*

$$\|Lf - f\|_\infty \leq \|f\|_\infty \cdot \|L1 - 1\|_\infty + \sum_{k=1}^{n} \frac{\left\|f^{(k)}\right\|_\infty}{k!} \left\|\left(L(t - \cdot)^k\right)(\cdot)\right\|_\infty$$

$$+ \omega_1\left(f^{(n)}, D_n\right) \cdot D_n^{n-1} \left(\frac{(M - m)}{(n + 1)!} + \frac{D_n}{2n!} + \frac{D_n^2}{8(M - m)(n - 1)!}\right). \tag{3.7}$$

By [1], p. 415, we have $D_n < +\infty$.

We derive

Theorem 3.2 *All as in Theorem 3.1. Then*

$$\|(Lf)(A) - f(A)\| \leq \|f(A)\| \, \|(L1)(A) - 1_H\|$$

$$+ \sum_{k=1}^{n} \frac{\left\| f^{(k)}(A) \right\|}{k!} \left\| \left(L(t-A)^k \right)(A) \right\|$$

$$+ \omega_1 \left(f^{(n)}, D_n \right) D_n^{n-1} \left(\frac{(M-m)}{(n+1)!} + \frac{D_n}{2n!} + \frac{D_n^2}{8(M-m)(n-1)!} \right), \tag{3.8}$$

where

$$D_n = \left\| \left(L\left(|t-A|^n \right) \right)(A) \right\|^{\frac{1}{n}}. \tag{3.9}$$

We mention

Corollary 3.3 ([1], p. 421) *Let L be a positive linear operator from* $C([m, M])$ *into itself. Here*

$$D_1 := \|(L(|t - \cdot|))(\cdot)\|_{\infty} < +\infty. \tag{3.10}$$

Let $f \in C^1([m, M])$. *Then*

$$\|Lf - f\|_{\infty} \leq \|f\|_{\infty} \cdot \|L1 - 1\|_{\infty} + \left\| f' \right\|_{\infty} \|(L(t - \cdot))(\cdot)\|_{\infty}$$

$$+ \frac{1}{2} \omega_1 \left(f', D_1 \right) \left((M-m) + D_1 + \frac{D_1^2}{4(M-m)} \right). \tag{3.11}$$

We obtain

Corollary 3.4 *All as in Corollary 3.3. Then*

$$\|(Lf)(A) - f(A)\| \leq \|f(A)\| \, \|(L1)(A) - 1_H\| + \left\| f'(A) \right\| \|(L(t-A))(A)\| \tag{3.12}$$

$$+ \frac{1}{2} \omega_1 \left(f', D_1 \right) \left((M-m) + D_1 + \frac{D_1^2}{4(M-m)} \right),$$

where

$$D_1 = \|(L(|t - A|))(A)\|. \tag{3.13}$$

We mention

Corollary 3.5 ([1], p. 421) *Let L be a positive linear operator from* $C([m, M])$ *into itself. Here*

$$D_2 := \left\| \left(L\left((t - \cdot)^2 \right) \right)(\cdot) \right\|_{\infty}^{\frac{1}{2}} < +\infty. \tag{3.14}$$

Let $f \in C^2([m, M])$. *Then*

$$\|Lf - f\|_\infty \leq \|f\|_\infty \cdot \|L1 - 1\|_\infty + \left\|f'\right\|_\infty \|(L(t - \cdot))(\cdot)\|_\infty$$

$$+ \frac{\left\|f''\right\|_\infty}{2} \left\|\left(L(t - \cdot)^2\right)(\cdot)\right\|_\infty$$

$$+ \frac{1}{2}\omega_1\left(f'', D_2\right) \cdot D_2 \left(\frac{(M - m)}{3} + \frac{D_2}{2} + \frac{D_2^2}{4(M - m)}\right). \tag{3.15}$$

We derive

Corollary 3.6 *All as in Corollary 3.5. Then*

$$\|(Lf)(A) - f(A)\| \leq \|f(A)\| \|(L1)(A) - 1_H\| + \left\|f'(A)\right\| \|(L(t - A))(A)\|$$

$$+ \frac{\left\|f''(A)\right\|}{2} \left\|\left(L(t - A)^2\right)(A)\right\|$$

$$+ \frac{1}{2}\omega_1\left(f'', D_2\right) D_2 \left(\frac{(M - m)}{3} + \frac{D_2}{2} + \frac{D_2^2}{4(M - m)}\right), \tag{3.16}$$

where

$$D_2 = \left\|\left(L\left((t - A)^2\right)\right)(A)\right\|^{\frac{1}{2}}. \tag{3.17}$$

We give

Example 3.7 Let $f \in C([0, 1])$, the basic Bernestein polynomial operators are defined by

$$(\beta_n(f))(t) := \sum_{k=0}^{n} f\left(\frac{k}{n}\right)\binom{n}{k} t^k (1 - t)^{n-k}, \quad t \in [0, 1]. \tag{3.18}$$

By [1], p. 421, when $f \in C^2([0, 1])$ we get

$$\|\beta_n(f) - f\|_\infty \leq \frac{\left\|f''\right\|_\infty}{8n} + \frac{1}{4\sqrt{n}}\omega_1\left(f'', \frac{1}{2\sqrt{n}}\right)\left(\frac{1}{3} + \frac{1}{4\sqrt{n}} + \frac{1}{16n}\right). \tag{3.19}$$

The map

$$[m, M] \ni s = \varphi(t) = (M - m)t + m, \quad t \in [0, 1], \tag{3.20}$$

maps $(1 - 1)$ and onto : $[0, 1]$ onto $[m, M]$.

Let now $f \in C^2([m, M])$, then

$$f(s) = f(\varphi(t)) = f((M - m)t + m), \tag{3.21}$$

and

$$\frac{df(s)}{dt} = (f(\varphi(t)))' = f'(\varphi(t))(M-m) = f'(s)(M-m). \tag{3.22}$$

Furthermore it holds

$$\frac{d^2 f(s)}{dt^2} = f''(s)(M-m)^2. \tag{3.23}$$

We observe that ($t \in [0,1]$)

$$(\beta_n(f((M-m)t+m)))(t) =$$

$$\sum_{k=0}^{n}\left(f\left((M-m)\frac{k}{n}+m\right)\right)\binom{n}{k} t^k (1-t)^{n-k} =$$

$$\sum_{k=0}^{n}\left(f\left((M-m)\frac{k}{n}+m\right)\right)\binom{n}{k}\left(\frac{s-m}{M-m}\right)^k\left(\frac{M-s}{M-m}\right)^{n-k} \tag{3.24}$$

$$=: (B_n(f))(s), \quad s \in [m,M].$$

The operators $(B_n(f))(s)$ are the general Bernstein polynomials. As in [5], we get that

$$\omega_1(f((M-m)t+m), \delta) = \omega_1(f, (M-m)\delta), \tag{3.25}$$

where $f \in C([m,M])$.

Here the function $f((M-m)t+m) \in C([0,1])$, as a function of $t \in [0,1]$. So we apply (3.19), for $f((M-m)t+m)$, $t \in [0,1]$, we obtain

$$\|B_n(f) - f\|_\infty \leq (M-m)^2\left[\frac{\|f''\|_\infty}{8n}\right. \tag{3.26}$$

$$\left.+\frac{1}{4\sqrt{n}}\omega_1\left(f'', \frac{(M-m)}{2\sqrt{n}}\right)\left(\frac{1}{3}+\frac{1}{4\sqrt{n}}+\frac{1}{16n}\right)\right],$$

where $f \in C([m,M])$.

Consequently, we obtain

$$\|(B_n(f))(A) - f(A)\| \leq (M-m)^2\left[\frac{\|f''(A)\|}{8n}\right. \tag{3.27}$$

$$\left.+\frac{1}{4\sqrt{n}}\omega_1\left(f'', \frac{(M-m)}{2\sqrt{n}}\right)\left(\frac{1}{3}+\frac{1}{4\sqrt{n}}+\frac{1}{16n}\right)\right],$$

$\forall f \in C([m,M])$.

We need

Theorem 3.8 ([1], p. 422) *Let $L \neq 0$ be a positive linear operator from $C([m, M])$ into itself. Set*

$$\rho := \left\| \left(L (t - x)^2 \right) (x) \right\|_{\infty}^{\frac{1}{2}}, \tag{3.28}$$

and consider $r > 0$. Let $f \in C^1([m, M])$. Then

$$\|Lf - f\|_{\infty} - \|f\|_{\infty} \|L1 - 1\|_{\infty} - \left\| f' \right\|_{\infty} \|(L(t - x))(x)\|_{\infty} \leq$$

$$\begin{cases} \frac{1}{8r} \left(2 + \sqrt{\|L(1)\|_{\infty}} r \right)^2 \omega_1 \left(f', r\rho \right) \rho, \ if \ r \leq \frac{2}{\sqrt{\|L(1)\|_{\infty}}}; \\ \sqrt{\|L(1)\|_{\infty}} \omega_1 \left(f', r\rho \right) \rho, \ if \ r > \frac{2}{\sqrt{\|L(1)\|_{\infty}}}, \end{cases} \tag{3.29}$$

by [1], p. 415, we have that $\rho < +\infty$.

An improved results for $f \in C^1([m, M])$ follows:

Theorem 3.9 *All as in Theorem 3.8. Then*

$$\|(Lf)(A) - f(A)\| - \|f(A)\| \|(L1)(A) - 1_H\| - \left\| f'(A) \right\| \|(L(t - A))(A)\| \leq \tag{3.30}$$

$$\begin{cases} \frac{1}{8r} \left(2 + \sqrt{\|(L(1))(A)\|} r \right)^2 \omega_1 \left(f', r\rho \right) \rho, \ if \ r \leq \frac{2}{\sqrt{\|(L(1))(A)\|}}; \\ \sqrt{\|(L(1))(A)\|} \omega_1 \left(f', r\rho \right) \rho, \ if \ r > \frac{2}{\sqrt{\|(L(1))(A)\|}}, \end{cases}$$

where

$$\rho = \left\| \left(L (t - A)^2 \right) (A) \right\|^{\frac{1}{2}}. \tag{3.31}$$

We continue with neural network operators.

Definition 3.10 (*see* [2], *pp. 3–12*) We consider here the sigmoidal function of logarithmic type

$$s(x) = \frac{1}{1 + e^{-x}}, \ x \in \mathbb{R}, \tag{3.32}$$

and

$$\Phi(x) = \frac{1}{2} (s(x + 1) - s(x - 1)) > 0, \ \forall \, x \in \mathbb{R}. \tag{3.33}$$

Let $f \in C([m, M])$ and $n \in \mathbb{N}$, such that $\lceil nm \rceil \leq \lfloor nM \rfloor$ ($\lceil \cdot \rceil$ is the ceiling and $\lfloor \cdot \rfloor$ is the integral part of the number).

We consider the positive linear neural network operator

$$G_n(f, x) = \frac{\sum_{k=\lceil nm \rceil}^{\lfloor nM \rfloor} f\left(\frac{k}{n}\right) \Phi(nx - k)}{\sum_{k=\lceil nm \rceil}^{\lfloor nM \rfloor} \Phi(nx - k)}, \ x \in [m, M]. \tag{3.34}$$

Clearly, $G_n : C([m, M]) \hookrightarrow C([m, M])$. For large enough n we always have $\lceil nm \rceil \leq \lfloor nM \rfloor$. Also $m \leq \frac{k}{n} \leq M$, iff $\lceil nm \rceil \leq k \leq \lfloor nM \rfloor$.

We need and mention

Theorem 3.11 (see [2], p. 9) *Let $f \in C([m, M])$, $0 < \alpha < 1$. Then*

$$\|G_n(f) - f\|_\infty \leq (5.250312578) \left[\omega_1 \left(f, \frac{1}{n^\alpha} \right) + 6.3984 \|f\|_\infty e^{-n^{(1-\alpha)}} \right]. \tag{3.35}$$

We derive

Theorem 3.12 *Let $f \in C([m, M])$, $0 < \alpha < 1$. Then*

$$\|(G_n(f))(A) - f(A)\| \leq (5.250312578) \left[\omega_1 \left(f, \frac{1}{n^\alpha} \right) + 6.3984 \|f(A)\| e^{-n^{(1-\alpha)}} \right]. \tag{3.36}$$

We mention

Theorem 3.13 ([2], p. 11) *Let $f \in C^N([m, M])$, $N \in \mathbb{N}$, $0 < \alpha < 1$. Then*

$$\|G_n(f) - f\|_\infty \leq (5.250312578) \cdot$$

$$\left\{ \sum_{j=1}^{N} \frac{\|f^{(j)}\|_\infty}{j!} \left[\frac{1}{n^{\alpha j}} + (3.1992)(M - m)^j e^{-n^{(1-\alpha)}} \right] \right.$$

$$\left. + \left[\omega_1 \left(f^{(N)}, \frac{1}{n^\alpha} \right) \frac{1}{n^{\alpha N} N!} + (6.3984) \frac{\|f^{(N)}\|_\infty}{N!} (M - m)^N e^{-n^{(1-\alpha)}} \right] \right\}. \tag{3.37}$$

To obtain

Theorem 3.14 *Let $f \in C^N([m, M])$, $N \in \mathbb{N}$, $0 < \alpha < 1$. Then*

$$\|(G_n(f))(A) - f(A)\| \leq (5.250312578) \cdot$$

$$\left\{ \sum_{j=1}^{N} \frac{\|f^{(j)}(A)\|}{j!} \left[\frac{1}{n^{\alpha j}} + (3.1992)(M - m)^j e^{-n^{(1-\alpha)}} \right] \right.$$

$$\left. + \left[\omega_1 \left(f^{(N)}, \frac{1}{n^\alpha} \right) \frac{1}{n^{\alpha N} N!} + (6.3984) \frac{\|f^{(N)}(A)\|}{N!} (M - m)^N e^{-n^{(1-\alpha)}} \right] \right\}. \tag{3.38}$$

We need

Definition 3.15 *([2], pp. 34–45)* We consider the hyperbolic tangent function $\tanh x$, $x \in \mathbb{R}$:

$$\tanh x := \frac{e^x - e^{-x}}{e^x + e^{-x}}, \tag{3.39}$$

and

$$\Psi(x) := \frac{1}{4}(\tanh(x+1) - \tanh(x-1)) > 0, \tag{3.40}$$

$\forall\, x \in \mathbb{R}$.

Let $f \in C([m, M])$ and $n \in \mathbb{N} : \lceil nm \rceil \leq \lfloor nM \rfloor$. We consider the positive linear neural network operator

$$F_n(f, x) = \frac{\sum_{k=\lceil nm \rceil}^{\lfloor nM \rfloor} f\left(\frac{k}{n}\right) \Psi(nx - k)}{\sum_{k=\lceil nm \rceil}^{\lfloor nM \rfloor} \Psi(nx - k)}, \quad x \in [m, M]. \tag{3.41}$$

Clearly, $F_n : C([m, M]) \hookrightarrow C([m, M])$.

We mention

Theorem 3.16 ([2], p. 42) *Let $f \in C([m, M])$, $0 < \alpha < 1$. Then*

$$\|F_n(f) - f\|_\infty \leq (4.1488766)\left[\omega_1\left(f, \frac{1}{n^\alpha}\right) + 2e^4 \|f\|_\infty e^{-2n^{(1-\alpha)}}\right]. \tag{3.42}$$

We derive

Theorem 3.17 *Let $f \in C([m, M])$, $0 < \alpha < 1$. Then*

$$\|(F_n(f))(A) - f(A)\| \leq (4.1488766)\left[\omega_1\left(f, \frac{1}{n^\alpha}\right) + 2e^4 \|f(A)\| e^{-2n^{(1-\alpha)}}\right]. \tag{3.43}$$

We mention

Theorem 3.18 ([2], p. 45) *Let $f \in C^N([m, M])$, $N \in \mathbb{N}$, $0 < \alpha < 1$. Then*

$$\|F_n(f) - f\|_\infty \leq (4.1488766) \cdot$$

$$\left\{\sum_{j=1}^{N}\frac{\|f^{(j)}\|_{\infty}}{j!}\left[\frac{1}{n^{\alpha j}}+e^{4}(M-m)^{j}e^{-2n^{(1-\alpha)}}\right]+\right. \tag{3.44}$$

$$\left.\left[\omega_1\left(f^{(N)},\frac{1}{n^{\alpha}}\right)\frac{1}{n^{\alpha N}N!}+\frac{2e^{4}\|f^{(N)}\|_{\infty}(M-m)^{N}}{N!}e^{-2n^{(1-\alpha)}}\right]\right\}.$$

We derive

Theorem 3.19 *Let $f \in C^{N}([m,M])$, $N \in \mathbb{N}$, $0<\alpha<1$. Then*

$$\|(F_n(f))(A)-f(A)\| \leq (4.1488766)\cdot$$

$$\left\{\sum_{j=1}^{N}\frac{\|f^{(j)}(A)\|}{j!}\left[\frac{1}{n^{\alpha j}}+e^{4}(M-m)^{j}e^{-2n^{(1-\alpha)}}\right]\right.$$

$$\left.+\left[\omega_1\left(f^{(N)},\frac{1}{n^{\alpha}}\right)\frac{1}{n^{\alpha N}N!}+\frac{2e^{4}\|f^{(N)}(A)\|(M-m)^{N}}{N!}e^{-2n^{(1-\alpha)}}\right]\right\}. \tag{3.45}$$

We make

Definition 3.20 (*[3], pp. 332–346*) We consider the (Gauss) error special function

$$erf(x)=\frac{2}{\sqrt{\pi}}\int_{0}^{x}e^{-t^{2}}dt,\quad x\in\mathbb{R}, \tag{3.46}$$

which is a sigmoidal type continuous function and it is a strictly increasing function.

We consider the activation function

$$\chi(x)=\frac{1}{4}(erf(x+1)-erf(x-1)),\quad x\in\mathbb{R}. \tag{3.47}$$

Notice $\chi(x)>0$, $\forall x\in\mathbb{R}$.

Let $f\in C([m,M])$, $n\in\mathbb{N}$ such that $n^{1-\alpha}\geq 3$, where $0<\alpha<1$.

We consider the positive linear operator

$$A_n(f,x)=\frac{\sum_{k=\lceil nm\rceil}^{\lfloor nM\rfloor}f\left(\frac{k}{n}\right)\chi(nx-k)}{\sum_{k=\lceil nm\rceil}^{\lfloor nM\rfloor}\chi(nx-k)},\quad \forall x\in[m,M]. \tag{3.48}$$

Th operator A_n is a neural network operator mapping $C([m,M])$ into itself.

We mention

Theorem 3.21 ([3], p. 340) *It holds*

$$\|A_n(f) - f\|_\infty \leq (4.019)\left[\omega_1\left(f, \frac{1}{n^\alpha}\right) + \frac{\|f\|_\infty}{\sqrt{\pi}\left(n^{1-\alpha}-2\right)e^{\left(n^{1-\alpha}-2\right)^2}}\right]. \quad (3.49)$$

We derive

Theorem 3.22 *It holds*

$$\|(A_n(f))(A) - f(A)\| \leq (4.019)\left[\omega_1\left(f, \frac{1}{n^\alpha}\right) + \frac{\|f(A)\|}{\sqrt{\pi}\left(n^{1-\alpha}-2\right)e^{\left(n^{1-\alpha}-2\right)^2}}\right]. \quad (3.50)$$

We need

Theorem 3.23 ([3], pp. 345–346) *Let $f \in C^N([m, M])$, $n, N \in \mathbb{N}$, $n^{1-\alpha} \geq 3$, $0 < \alpha < 1$. Then*

$$\|A_n(f) - f\|_\infty \leq (4.019)\cdot$$

$$\left\{\sum_{j=1}^{N}\frac{\left\|f^{(j)}\right\|_\infty}{j!}\left[\frac{1}{n^{\alpha j}} + \frac{(M-m)^j}{2\sqrt{\pi}\left(n^{1-\alpha}-2\right)e^{\left(n^{1-\alpha}-2\right)^2}}\right]\right.$$

$$\left.+\left[\omega_1\left(f^{(N)}, \frac{1}{n^\alpha}\right)\frac{1}{n^{\alpha N}N!} + \frac{\left\|f^{(N)}\right\|_\infty (M-m)^N}{N!\sqrt{\pi}\left(n^{1-\alpha}-2\right)e^{\left(n^{1-\alpha}-2\right)^2}}\right]\right\}. \quad (3.51)$$

It follows

Theorem 3.24 *All as in Theorem 3.23. Then*

$$\|(A_n(f))(A) - f(A)\| \leq (4.019)\cdot$$

$$\left\{\sum_{j=1}^{N}\frac{\left\|f^{(j)}(A)\right\|}{j!}\left[\frac{1}{n^{\alpha j}} + \frac{(M-m)^j}{2\sqrt{\pi}\left(n^{1-\alpha}-2\right)e^{\left(n^{1-\alpha}-2\right)^2}}\right]\right.$$

$$\left.+\left[\omega_1\left(f^{(N)}, \frac{1}{n^\alpha}\right)\frac{1}{n^{\alpha N}N!} + \frac{\left\|f^{(N)}(A)\right\| (M-m)^N}{N!\sqrt{\pi}\left(n^{1-\alpha}-2\right)e^{\left(n^{1-\alpha}-2\right)^2}}\right]\right\}. \quad (3.52)$$

Conclusion 3.25 *Inequalities (3.8), (3.12), (3.16), (3.27), (3.30), (3.36), (3.38), (3.43), (3.45), (3.50) and (3.52), imply $\|(Lf)(A) - f(A)\| \to 0$, under basic assumptions and imply $\|(B_n(f))(A) - f(A)\| \to 0$, $\|(G_n(f))(A) - f(A)\| \to 0$, $\|(F_n(f))(A) - f(A)\| \to 0$, and $\|(A_n(f))(A) - f(A)\| \to 0$, as $n \to \infty$.*

The approximations are given quantitatively and with rates via the first modulus of continuity.

References

1. G. Anastassiou, *Quantitative Approximation* (Chapman & Hall / CRC, Boca Raton, New York, 2001)
2. G. Anastassiou, *Intelligent Systems: Approximation by Artificial Neural Networks* (Springer, Heidelberg, New York, 2011)
3. G. Anastassiou, *Intelligent Systems II: Complete Approximation by Neural Network Operators* (Springer, Heidelberg, 2016)
4. G. Anastassiou, *Quantitative Self Adjoint Operator Direct Approximations*, J. Nonlinear Sci. Appl. Accepted, 2016
5. G. Anastassiou, *Self Adjoint Operator Korovkin Type and Polynomial Direct Approximations with Rates* (2016)
6. S.S. Dragomir, *Inequalities for Functions of Selfadjoint Operators on Hilbert Spaces* (2011), ajmaa.org/RGMIA/monographs/InFuncOp.pdf
7. S. Dragomir, *Operator inequalities of Ostrowski and Trapezoidal type* (Springer, New York, 2012)
8. T. Furuta, J. Mićić Hot, J. Pečaric, Y. Seo, *Mond-Pečaric Method in Operator Inequalities. Inequalities for Bounded Selfadjoint Operators on a Hilbert Space*, Element, Zagreb (2005)
9. G. Helmberg, *Introduction to Spectral Theory in Hilbert Space* (Wiley, New York, 1969)
10. C.A. McCarthy, c_p. Isr. J. Math. **5**, 249–271 (1967)

Chapter 4
Fractional Self Adjoint Operator Poincaré and Sobolev Inequalities

We present here many fractional self adjoint operator Poincaré and Sobolev type inequalities to various directions. Initially we give several fractional representation formulae in the self adjoint operator sense. Inequalities are based in the self adjoint operator order over a Hilbert space. It follows [3].

4.1 Background

Let A be a selfadjoint linear operator on a complex Hilbert space $(H; \langle \cdot, \cdot \rangle)$. The Gelfand map establishes a $*$–isometrically isomorphism Φ between the set $C(Sp(A))$ of all continuous functions defined on the spectrum of A, denoted $Sp(A)$, and the C^*-algebra $C^*(A)$ generated by A and the identity operator 1_H on H as follows (see e.g. [7, p. 3]):

For any $f, g \in C(Sp(A))$ and any $\alpha, \beta \in \mathbb{C}$ we have

(i) $\Phi(\alpha f + \beta g) = \alpha \Phi(f) + \beta \Phi(g)$;
(ii) $\Phi(fg) = \Phi(f) \Phi(g)$ (the operation composition is on the right) and $\Phi(\overline{f}) = (\Phi(f))^*$;
(iii) $\|\Phi(f)\| = \|f\| := \sup_{t \in Sp(A)} |f(t)|$;
(iv) $\Phi(f_0) = 1_H$ and $\Phi(f_1) = A$, where $f_0(t) = 1$ and $f_1(t) = t$, for $t \in Sp(A)$.

With this notation we define

$$f(A) := \Phi(f), \text{ for all } f \in C(Sp(A)),$$

and we call it the continuous functional calculus for a selfadjoint operator A.

If A is a selfadjoint operator and f is a real valued continuous function on $Sp(A)$ then $f(t) \geq 0$ for any $t \in Sp(A)$ implies that $f(A) \geq 0$, i.e. $f(A)$ is a positive

G.A. Anastassiou, *Intelligent Comparisons II: Operator Inequalities and Approximations*, Studies in Computational Intelligence 699,
DOI 10.1007/978-3-319-51475-8_4

operator on H. Moreover, if both f and g are real valued continuous functions on $Sp(A)$ then the following important property holds:

(P) $f(t) \geq g(t)$ for any $t \in Sp(A)$, implies that $f(A) \geq g(A)$ in the operator order of $B(H)$ (the Banach algebra of all bounded linear operators from H into itself).

Equivalently, we use (see [6], pp. 7–8):

Let U be a selfadjoint operator on the complex Hilbert space $(H, \langle \cdot, \cdot \rangle)$ with the spectrum $Sp(U)$ included in the interval $[m, M]$ for some real numbers $m < M$ and $\{E_\lambda\}_\lambda$ be its spectral family.

Then for any continuous function $f : [m, M] \to \mathbb{C}$, it is well known that we have the following spectral representation in terms of the Riemann–Stieljes integral:

$$\langle f(U)x, y \rangle = \int_{m-0}^{M} f(\lambda) \, d(\langle E_\lambda x, y \rangle),$$

for any $x, y \in H$. The function $g_{x,y}(\lambda) := \langle E_\lambda x, y \rangle$ is of bounded variation on the interval $[m, M]$, and

$$g_{x,y}(m-0) = 0 \text{ and } g_{x,y}(M) = \langle x, y \rangle,$$

for any $x, y \in H$. Furthermore, it is known that $g_x(\lambda) := \langle E_\lambda x, x \rangle$ is increasing and right continuous on $[m, M]$.

We have also the formula

$$\langle f(U)x, x \rangle = \int_{m-0}^{M} f(\lambda) \, d(\langle E_\lambda x, x \rangle), \quad \forall x \in H.$$

As a symbol we can write

$$f(U) = \int_{m-0}^{M} f(\lambda) \, dE_\lambda.$$

Above, $m = \min\{\lambda | \lambda \in Sp(U)\} := \min Sp(U)$, $M = \max\{\lambda | \lambda \in Sp(U)\} := \max Sp(U)$. The projections $\{E_\lambda\}_{\lambda \in \mathbb{R}}$, are called the spectral family of A, with the properties:

(a) $E_\lambda \leq E_{\lambda'}$ for $\lambda \leq \lambda'$;

(b) $E_{m-0} = 0_H$ (zero operator), $E_M = 1_H$ (identity operator) and $E_{\lambda+0} = E_\lambda$ for all $\lambda \in \mathbb{R}$.

Furthermore

$$E_\lambda := \varphi_\lambda(U) \ \forall \lambda \in \mathbb{R},$$

is a projection which reduces U, with

$$\varphi_\lambda(s) := \begin{cases} 1, \text{ for } -\infty < s \leq \lambda, \\ 0, \text{ for } \lambda < s < +\infty. \end{cases}$$

The spectral family $\{E_\lambda\}_{\lambda\in\mathbb{R}}$ determines uniquely the self-adjoint operator U and vice versa.

For more on the topic see [8], pp. 256–266, and for more details see there pp. 157–266. See also [5].

Some more basics are given (we follow [6], pp. 1–5):

Let $(H; \langle\cdot,\cdot\rangle)$ be a Hilbert space over $\mathbb{C}$. A bounded linear operator A defined on H is selfjoint, i.e., $A = A^*$, iff $\langle Ax, x\rangle \in \mathbb{R}$, $\forall\, x \in H$, and if A is selfadjoint, then

$$\|A\| = \sup_{x\in H:\|x\|=1} |\langle Ax, x\rangle|.$$

Let A, B be selfadjoint operators on H. Then $A \le B$ iff $\langle Ax, x\rangle \le \langle Bx, x\rangle, \forall\, x \in H$.

In particular, A is called positive if $A \ge 0$.

Denote by

$$\mathcal{P} := \left\{\varphi(s) := \sum_{k=0}^{n} \alpha_k s^k | n \ge 0, \alpha_k \in \mathbb{C}, 0 \le k \le n\right\}.$$

If $A \in \mathcal{B}(H)$ is selfadjoint, and $\varphi(s) \in \mathcal{P}$ has real coefficients, then $\varphi(A)$ is selfadjoint, and

$$\|\varphi(A)\| = \max\{|\varphi(\lambda)|, \lambda \in Sp(A)\}.$$

If φ is any function defined on $\mathbb{R}$ we define

$$\|\varphi\|_A := \sup\{|\varphi(\lambda)|, \lambda \in Sp(A)\}.$$

If A is selfadjoint operator on Hilbert space H and φ is continuous and given that $\varphi(A)$ is selfadjoint, then $\|\varphi(A)\| = \|\varphi\|_A$. And if φ is a continuous real valued function so it is $|\varphi|$, then $\varphi(A)$ and $|\varphi|(A) = |\varphi(A)|$ are selfadjoint operators (by [6], p. 4, Theorem 7).

Hence it holds

$$\begin{aligned} \||\varphi(A)|\| &= \||\varphi|\|_A = \sup\{||\varphi(\lambda)||, \lambda \in Sp(A)\} \\ &= \sup\{|\varphi(\lambda)|, \lambda \in Sp(A)\} = \|\varphi\|_A = \|\varphi(A)\|, \end{aligned}$$

that is

$$\||\varphi(A)|\| = \|\varphi(A)\|.$$

For a selfadjoint operator $A \in \mathcal{B}(H)$ which is positive, there exists a unique positive selfadjoint operator $B := \sqrt{A} \in \mathcal{B}(H)$ such that $B^2 = A$, that is $\left(\sqrt{A}\right)^2 = A$. We call B the square root of A.

Let $A \in \mathcal{B}(H)$, then A^*A is selfadjoint and positive. Define the "operator absolute value" $|A| := \sqrt{A^*A}$. If $A = A^*$, then $|A| = \sqrt{A^2}$.

For a continuous real valued function φ we observe the following:

$$|\varphi(A)| \text{ (the functional absolute value) } = \int_{m-0}^{M} |\varphi(\lambda)|\, dE_{\lambda} =$$

$$\int_{m-0}^{M} \sqrt{(\varphi(\lambda))^2} dE_{\lambda} = \sqrt{(\varphi(A))^2} = |\varphi(A)| \text{ (operator absolute value)},$$

where A is a selfadjoint operator.

That is we have

$$|\varphi(A)| \text{ (functional absolute value) } = |\varphi(A)| \text{ (operator absolute value)}.$$

4.2 Main Results

Let A be a selfadjoint operator in the Hilbert space H with the spectrum $Sp(A) \subseteq [m, M]$, $m < M$; $m, M \in \mathbb{R}$.

In the next we obtain fractional Poincaré and Sobolev type inequalities in the operator order of $\mathcal{B}(H)$ (the Banach algebra of all bounded linear operators from H into itself). All of our functions next in this chapter are real valued.

We give

Definition 4.1 ([1], *p. 270*) Let $\nu > 0$, $n := \lceil \nu \rceil$ (ceiling of ν), $f \in AC^n([m, M])$ (i.e. $f^{(n-1)}$ is absolutely continuous on $[m, M]$, that is in $AC([m, M])$). We define the left Caputo fractional derivative

$$\left(D_{*m}^{\nu} f\right)(z) := \frac{1}{\Gamma(n-\nu)} \int_{m}^{z} (z-t)^{n-\nu-1} f^{(n)}(t)\, dt, \tag{4.1}$$

which exists almost everywhere for $z \in [m, M]$.

Notice that $D_{*m}^{0} f = f$, and $D_{*m}^{n} f = f^{(n)}$.

We present the operator representation formula.

Theorem 4.2 *Let A be a selfadjoint operator in the Hilbert space H with the spectrum $Sp(A) \subseteq [m, M]$ for some real numbers $m < M$, $\{E_{\lambda}\}_{\lambda}$ be its spectral family, I be a closed subinterval on $\mathbb{R}$ with $[m, M] \subset \mathring{I}$ (the interior of I) and $n \in \mathbb{N}$, with $n := \lceil \nu \rceil$, $\nu > 0$. We consider $f \in AC^n([m, M])$ (i.e. $f^{(n-1)} \in AC([m, M])$, absolutely continuous functions), where $f : I \to \mathbb{R}$.*

Then

$$f(A) = \sum_{k=0}^{n-1} \frac{f^{(k)}(m)}{k!} (A - m1_H)^k + R_n(f, m, M), \tag{4.2}$$

where

$$R_n(f, m, M) = \frac{1}{\Gamma(\nu)} \int_{m-0}^{M} \left(\int_m^{\lambda} (\lambda - t)^{\nu-1} \left(D_{*m}^{\nu} f \right)(t) \, dt \right) dE_{\lambda}. \quad (4.3)$$

Proof We have by left Caputo fractional Taylor's formula [4], p. 54, that

$$f(\lambda) = \sum_{k=0}^{n-1} \frac{f^{(k)}(m)}{k!} (\lambda - m)^k + \frac{1}{\Gamma(\nu)} \int_m^{\lambda} (\lambda - t)^{\nu-1} \left(D_{*m}^{\nu} f \right)(t) \, dt, \quad (4.4)$$

$\forall \lambda \in [m, M]$.

Then we integrate (4.4) against E_{λ} to get

$$\int_{m-0}^{M} f(\lambda) \, dE_{\lambda} = \sum_{k=0}^{n-1} \frac{f^{(k)}(m)}{k!} \int_{m-0}^{M} (\lambda - m)^k \, dE_{\lambda} +$$

$$\frac{1}{\Gamma(\nu)} \int_{m-0}^{M} \left(\int_m^{\lambda} (\lambda - t)^{\nu-1} \left(D_{*m}^{\nu} f \right)(t) \, dt \right) dE_{\lambda}. \quad (4.5)$$

By the spectral representation theorem we obtain

$$f(A) = \sum_{k=0}^{n-1} \frac{f^{(k)}(m)}{k!} (A - m 1_H)^k + \quad (4.6)$$

$$\frac{1}{\Gamma(\nu)} \int_{m-0}^{M} \left(\int_m^{\lambda} (\lambda - t)^{\nu-1} \left(D_{*m}^{\nu} f \right)(t) \, dt \right) dE_{\lambda},$$

proving the claim. ■

Remark 4.3 In (4.6) assume that $f^{(k)}(m) = 0$, $k = 0, \ldots, n-1$. Then

$$f(A) = \frac{1}{\Gamma(\nu)} \int_{m-0}^{M} \left(\int_m^{\lambda} (\lambda - t)^{\nu-1} \left(D_{*m}^{\nu} f \right)(t) \, dt \right) dE_{\lambda}. \quad (4.7)$$

Therefore it holds

$$\langle f(A) x, y \rangle = \frac{1}{\Gamma(\nu)} \int_{m-0}^{M} \left(\int_m^{\lambda} (\lambda - t)^{\nu-1} \left(D_{*m}^{\nu} f \right)(t) \, dt \right) d \langle E_{\lambda} x, y \rangle, \quad (4.8)$$

$\forall x, y \in H$.

The function $g_{x,y}(\lambda) := \langle E_{\lambda} x, y \rangle$ is of bounded variation on the interval $[m, M]$ and

$$g_{x,y}(m - 0) = 0 \text{ and } g_{x,y}(M) = \langle x, y \rangle, \ \forall x, y \in H. \quad (4.9)$$

It is also well known that $g_x(\lambda) := \langle E_\lambda x, x\rangle$ is nondecreasing and right continuous on $[m, M]$.

One has

$$\langle f(A)x, x\rangle = \frac{1}{\Gamma(\nu)}\int_{m-0}^{M}\left(\int_m^\lambda (\lambda-t)^{\nu-1}\left(D_{*m}^\nu f\right)(t)\,dt\right)d\langle E_\lambda x, x\rangle, \quad (4.10)$$

$\forall x \in H$.

Remark 4.4 (*all as in Theorem* 4.2, *Remark* 4.3) Let $p, q > 1 : \frac{1}{p} + \frac{1}{q} = 1$, with $\nu > \frac{1}{q}$. Then

$$\int_m^\lambda (\lambda-t)^{\nu-1}\left|\left(D_{*m}^\nu f\right)(t)\right|dt \leq$$

$$\left(\int_m^\lambda (\lambda-t)^{p(\nu-1)}\,dt\right)^{\frac{1}{p}}\left(\int_m^\lambda \left|\left(D_{*m}^\nu f\right)(t)\right|^q dt\right)^{\frac{1}{q}} \leq \quad (4.11)$$

$$\frac{(\lambda-m)^{\frac{p(\nu-1)+1}{p}}}{(p(\nu-1)+1)^{\frac{1}{p}}}\left(\int_m^M \left|\left(D_{*m}^\nu f\right)(t)\right|^q dt\right)^{\frac{1}{q}} =$$

$$\frac{(\lambda-m)^{\nu-1+\frac{1}{p}}}{(p(\nu-1)+1)^{\frac{1}{p}}}\left\|D_{*m}^\nu f\right\|_{q,[m,M]} = \frac{(\lambda-m)^{\nu-\frac{1}{q}}}{(p(\nu-1)+1)^{\frac{1}{p}}}\left\|D_{*m}^\nu f\right\|_{q,[m,M]}. \quad (4.12)$$

We have proved that

$$\left|\int_m^\lambda (\lambda-t)^{\nu-1}\left(D_{*m}^\nu f\right)(t)\,dt\right| \leq \int_m^\lambda (\lambda-t)^{\nu-1}\left|\left(D_{*m}^\nu f\right)(t)\right|dt \leq$$

$$\frac{(\lambda-m)^{\nu-\frac{1}{q}}}{(p(\nu-1)+1)^{\frac{1}{p}}}\left\|D_{*m}^\nu f\right\|_{q,[m,M]}, \quad (4.13)$$

$\forall \lambda \in [m, M]$.

Therefore it holds

$$|\langle f(A)x, x\rangle| \overset{(4.10)}{\leq} \frac{1}{\Gamma(\nu)}\int_{m-0}^{M}\left(\int_m^\lambda (\lambda-t)^{\nu-1}\left(D_{*m}^\nu f\right)(t)\,dt\right)d\langle E_\lambda x, x\rangle \quad (4.14)$$

$$\leq \frac{\left\|D_{*m}^\nu f\right\|_{q,[m,M]}}{(p(\nu-1)+1)^{\frac{1}{p}}\Gamma(\nu)}\int_{m-0}^{M}(\lambda-m)^{\nu-\frac{1}{q}}\,d\langle E_\lambda x, x\rangle =$$

$$\frac{\left\|D_{*m}^\nu f\right\|_{q,[m,M]}}{(p(\nu-1)+1)^{\frac{1}{p}}\Gamma(\nu)}\left\langle (A-m1_H)^{\nu-\frac{1}{q}}x, x\right\rangle, \quad \forall x \in H.$$

We have proved

Theorem 4.5 *All as in Theorem 4.2. Assume further $f^{(k)}(m) = 0$, $k = 0, 1, \dots, n-1$. Let $p, q > 1 : \frac{1}{p} + \frac{1}{q} = 1$, with $\nu > \frac{1}{q}$. Then*

$$|\langle f(A)x, x\rangle| \leq \frac{\left\| D_{*m}^{\nu} f \right\|_{q,[m,M]}}{(p(\nu-1)+1)^{\frac{1}{p}} \Gamma(\nu)} \left\langle (A - m1_H)^{\nu - \frac{1}{q}} x, x \right\rangle, \tag{4.15}$$

$\forall x \in H$.

Inequality (4.15) means that

$$\|f(A)\| \leq \frac{\left\| D_{*0m}^{\nu} f \right\|_{q,[m,M]}}{(p(\nu-1)+1)^{\frac{1}{p}} \Gamma(\nu)} \left\| (A - m1_H)^{\nu - \frac{1}{q}} \right\| \tag{4.16}$$

and in particular,

$$f(A) \leq \frac{\left\| D_{*m}^{\nu} f \right\|_{q,[m,M]}}{(p(\nu-1)+1)^{\frac{1}{p}} \Gamma(\nu)} (A - m1_H)^{\nu - \frac{1}{q}}. \tag{4.17}$$

We need

Definition 4.6 Let the real valued function $f \in C([m, M])$, and we consider

$$g(t) = \int_m^t f(z)\, dz \,\forall\, t \in [m, M], \tag{4.18}$$

then $g \in C([m, M])$.

We denote by

$$\int_{m1_H}^{A} f := \Phi(g) = g(A). \tag{4.19}$$

We understand and write that $(r > 0)$

$$g^r(A) = \Phi(g^r) =: \left(\int_{m1_H}^{A} f \right)^r.$$

Clearly $\left(\int_{m1_H}^{A} f \right)^r$ is a self adjoint operator on H, for any $r > 0$.

We will use

Theorem 4.7 ([1], p. 451) *Let $\nu \geq \gamma + 1$, $\gamma \geq 0$, $n := \lceil \nu \rceil$ ($\lceil \cdot \rceil$ ceiling of number). Assume $f \in C^n([m, M])$ such that $f^{(k)}(m) = 0$, $k = 0, 1, \dots, n-1$. Let $p, q > 1 : \frac{1}{p} + \frac{1}{q} = 1$. Then*

$$\int_m^{\lambda} \left|D_{*m}^{\gamma} f(t)\right|^q dt \leq \tag{4.20}$$

$$\left[\frac{(\lambda - m)^{q(\nu-\gamma)}}{(\Gamma(\nu-\gamma))^q (p(\nu-\gamma-1)+1)^{\frac{q}{p}} q(\nu-\gamma)}\right] \int_m^{\lambda} \left|D_{*m}^{\nu} f(t)\right|^q dt,$$

$\forall \lambda \in [m, M]$.

Note: By Proposition 15.114 ([1], p. 388) we have that $D_{*m}^{\nu} f, D_{*m}^{\gamma} f \in C([m, M])$. Using (4.20) and properties (P) and (ii), we derive the operator Poincaré inequality:

Theorem 4.8 *All as in Theorem 4.7. Then*

$$\int_{m1_H}^{A} \left|D_{*m}^{\gamma} f\right|^q \leq$$

$$\left[\frac{(A - m1_H)^{q(\nu-\gamma)}}{(\Gamma(\nu-\gamma))^q (p(\nu-\gamma-1)+1)^{\frac{q}{p}} q(\nu-\gamma)}\right] \left(\int_{m1_H}^{A} \left|D_{*m}^{\nu} f\right|^q\right). \tag{4.21}$$

We will use

Theorem 4.9 ([1], p. 493) *Let* $\nu \geq \gamma + 1, \gamma \geq 0, n := \lceil \nu \rceil$. *Assume* $f \in C^n([m, M])$ *such that* $f^{(k)}(m) = 0, k = 0, 1, \ldots, n-1$. *Let* $p, q > 1 : \frac{1}{p} + \frac{1}{q} = 1, r \geq 1$. *Then*

$$\left(\int_m^{\lambda} \left|D_{*m}^{\gamma} f(t)\right|^r dt\right)^{\frac{1}{r}} \leq$$

$$\left[\frac{(\lambda - m)^{\nu-\gamma+\frac{1}{r}-\frac{1}{q}}}{(\Gamma(\nu-\gamma))(p(\nu-\gamma-1)+1)^{\frac{1}{p}}}\right] \frac{\left(\int_m^{\lambda} \left|D_{*m}^{\nu} f(t)\right|^q dt\right)^{\frac{1}{q}}}{\left[r\left(\nu-\gamma-\frac{1}{q}\right)+1\right]^{\frac{1}{r}}}, \tag{4.22}$$

$\forall \lambda \in [m, M]$.

Applying (4.22), using properties (P) and (ii), we get the following operator Sobolev type inequality:

Theorem 4.10 *All as in Theorem 4.9. Then*

$$\left(\int_{m1_H}^{A} \left|D_{*m}^{\gamma} f\right|^r\right)^{\frac{1}{r}} \leq$$

$$\frac{(A - m1_H)^{\nu-\gamma+\frac{1}{r}-\frac{1}{q}}}{(\Gamma(\nu-\gamma))(p(\nu-\gamma-1)+1)^{\frac{1}{p}}} \frac{\left(\int_{m1_H}^{A} \left|D_{*m}^{\nu} f\right|^q\right)^{\frac{1}{q}}}{\left[r\left(\nu-\gamma-\frac{1}{q}\right)+1\right]^{\frac{1}{r}}}. \tag{4.23}$$

Next we follow [1], p. 8.

Definition 4.11 Let $\nu > 0$, $n := [\nu]$ (integral part), and $\alpha := \nu - n$ $(0 < \alpha < 1)$. Let $f \in C([m, M])$ and define

$$\left(J_{\nu}^{m} f\right)(z) = \frac{1}{\Gamma(\nu)} \int_{m}^{z} (z - t)^{\nu-1} f(t)\, dt, \tag{4.24}$$

all $m \leq z \leq M$, where Γ is the gamma function, the left generalized Riemann–Liouville integral. We define the subspace $C_m^{\nu}([m, M])$ of $C^n([m, M])$:

$$C_m^{\nu}([m, M]) := \left\{ f \in C^n([m, M]) : J_{1-\alpha}^{m} f^{(n)} \in C^1([m, M]) \right\}. \tag{4.25}$$

So let $f \in C_m^{\nu}([m, M])$; we define the left generalized ν-fractional derivative (of Canavati type) of f over $[m, M]$ as

$$D_m^{\nu} f := \left(J_{1-\alpha}^{m} f^{(n)}\right)'. \tag{4.26}$$

Notice that

$$\left(J_{1-\alpha}^{m} f^{(n)}\right)(z) = \frac{1}{\Gamma(1-\alpha)} \int_{m}^{z} (z - t)^{-\alpha} f^{(n)}(t)\, dt \tag{4.27}$$

exists for $f \in C_m^{\nu}([m, M])$, all $m \leq z \leq M$.

Also we notice that $D_m^{\nu} f \in C([m, M])$, $D_m^n f = f^{(n)}$, $n \in \mathbb{N}$; $D_m^0 f = f$.

We need

Theorem 4.12 ([1], p. 9) *Let $f \in C_m^{\nu}([m, M])$. Then*

(i) for $\nu \geq 1$, we have

$$f(\lambda) = f(m) + f'(m)(\lambda - m) + \frac{f''(m)}{2}(\lambda - m)^2 + \ldots + \tag{4.28}$$

$$f^{(n-1)}(m) \frac{(\lambda - m)^{n-1}}{(n-1)!} + \frac{1}{\Gamma(\nu)} \int_{m}^{\lambda} (\lambda - t)^{\nu-1} \left(D_m^{\nu} f\right)(t)\, dt,$$

(ii) if $0 < \nu < 1$ we get

$$f(\lambda) = \frac{1}{\Gamma(\nu)} \int_{m}^{\lambda} (\lambda - t)^{\nu-1} \left(D_m^{\nu} f\right)(t)\, dt, \tag{4.29}$$

$\forall\, \lambda \in [m, M]$.

We present the following operator representation formula:

Theorem 4.13 *Let A be a selfadjoint operator in the Hilbert space H with the spectrum $Sp(A) \subseteq [m, M]$ for some real numbers $m < M$, $\{E_{\lambda}\}_{\lambda}$ be its spectral family, $[m, M] \subset (a, b)$ and $n \in \mathbb{N}$, where $n := \lceil \nu \rceil$, $\nu > 0$. We consider $f \in C_m^{\nu}([m, M])$, where $f : [a, b] \rightarrow \mathbb{R}$.*

Then

(i) *for* $\nu \geq 1$, *we have*

$$f(A) = \sum_{k=0}^{n-1} \frac{f^{(k)}(m)}{k!} (A - m1_H)^k + R_n^*(f, m, M), \tag{4.30}$$

where

$$R_n^*(f, m, M) = \frac{1}{\Gamma(\nu)} \int_{m-0}^{M} \left(\int_m^{\lambda} (\lambda - t)^{\nu-1} \left(D_m^{\nu} f\right)(t)\, dt \right) dE_{\lambda}. \tag{4.31}$$

(ii) *if* $0 < \nu < 1$ *we get*

$$f(A) = \frac{1}{\Gamma(\nu)} \int_{m-0}^{M} \left(\int_m^{\lambda} (\lambda - t)^{\nu-1} \left(D_m^{\nu} f\right)(t)\, dt \right) dE_{\lambda}. \tag{4.32}$$

Proof We integrate (4.28), (4.29) against E_{λ}, apply spectral representation theorem. ■

Remark 4.14 In (4.30) ($\nu \geq 1$) we assume $f^{(k)}(m) = 0$, $k = 0, 1, \ldots, n-1$, then

$$f(A) = \frac{1}{\Gamma(\nu)} \int_{m-0}^{M} \left(\int_m^{\lambda} (\lambda - t)^{\nu-1} \left(D_m^{\nu} f\right)(t)\, dt \right) dE_{\lambda}. \tag{4.33}$$

We have

$$\langle f(A)x, x \rangle = \frac{1}{\Gamma(\nu)} \int_{m-0}^{M} \left(\int_m^{\lambda} (\lambda - t)^{\nu-1} \left(D_m^{\nu} f\right)(t)\, dt \right) d\langle E_{\lambda} x, x \rangle, \tag{4.34}$$

$\forall\, x \in H$.

Let $p, q > 1 : \frac{1}{p} + \frac{1}{q} = 1$, with $\nu > \frac{1}{q}$. Then

$$\left| \int_m^{\lambda} (\lambda - t)^{\nu-1} \left(D_m^{\nu} f\right)(t)\, dt \right| \leq \int_m^{\lambda} (\lambda - t)^{\nu-1} \left| \left(D_m^{\nu} f\right)(t) \right| dt \leq$$

$$\frac{(\lambda - m)^{\nu - \frac{1}{q}}}{(p(\nu - 1) + 1)^{\frac{1}{p}}} \left\| D_m^{\nu} f \right\|_{q,[m,M]}, \tag{4.35}$$

$\forall\, \lambda \in [m, M]$.

Hence

$$|\langle f(A)x,x\rangle| \overset{(4.34)}{\leq} \frac{1}{\Gamma(\nu)}\int_{m-0}^{M}\left|\int_{m}^{\lambda}(\lambda-t)^{\nu-1}\left(D_{m}^{\nu}f\right)(t)\,dt\right|d\langle E_{\lambda}x,x\rangle$$

$$\leq \frac{\left\|D_{m}^{\nu}f\right\|_{q,[m,M]}}{(p(\nu-1)+1)^{\frac{1}{p}}\Gamma(\nu)}\int_{m-0}^{M}(\lambda-m)^{\nu-\frac{1}{q}}\,d\langle E_{\lambda}x,x\rangle = \tag{4.36}$$

$$\frac{\left\|D_{m}^{\nu}f\right\|_{q,[m,M]}}{(p(\nu-1)+1)^{\frac{1}{p}}\Gamma(\nu)}\left\langle (A-m1_{H})^{\nu-\frac{1}{q}}x,x\right\rangle,\ \forall x\in H.$$

We have proved

Theorem 4.15 *All as in Theorem 4.13. Let $\nu > 0$. In case of $\nu \geq 1$, assume further $f^{(k)}(m) = 0$, $k = 0, 1, \ldots, n-1$. Let $p, q > 1 : \frac{1}{p} + \frac{1}{q} = 1$, with $\nu > \frac{1}{q}$. Then*

$$|\langle f(A)x,x\rangle| \leq \frac{\left\|D_{m}^{\nu}f\right\|_{q,[m,M]}}{(p(\nu-1)+1)^{\frac{1}{p}}\Gamma(\nu)}\left\langle (A-m1_{H})^{\nu-\frac{1}{q}}x,x\right\rangle, \tag{4.37}$$

$\forall x \in H$.

Inequality (4.37) means that

$$\|f(A)\| \leq \frac{\left\|D_{m}^{\nu}f\right\|_{q,[m,M]}}{(p(\nu-1)+1)^{\frac{1}{p}}\Gamma(\nu)}\left\|(A-m1_{H})^{\nu-\frac{1}{q}}\right\|, \tag{4.38}$$

and in particular,

$$f(A) \leq \left(\frac{\left\|D_{m}^{\nu}f\right\|_{q,[m,M]}}{(p(\nu-1)+1)^{\frac{1}{p}}\Gamma(\nu)}\right)(A-m1_{H})^{\nu-\frac{1}{q}}. \tag{4.39}$$

We will use

Theorem 4.16 ([1], p. 447) *Let $\nu \geq \gamma + 1$, $\gamma \geq 0$, $n := [\nu]$. Assume $f \in C_{m}^{\nu}([m, M])$ such that $f^{(k)}(m) = 0$, $k = 0, 1, \ldots, n-1$. Let $p, q > 1 : \frac{1}{p} + \frac{1}{q} = 1$. Then*

$$\int_{m}^{\lambda}\left|D_{m}^{\gamma}f(t)\right|^{q}dt \leq \tag{4.40}$$

$$\left[\frac{(\lambda-m)^{q(\nu-\gamma)}}{(\Gamma(\nu-\gamma))^{q}(p(\nu-\gamma-1)+1)^{\frac{q}{p}}q(\nu-\gamma)}\right]\int_{m}^{\lambda}\left|D_{m}^{\nu}f(t)\right|^{q}dt,$$

$\forall\, \lambda \in [m, M]$.

By Remark 3.4, [1], p. 26, $D_{m}^{\gamma}f \in C([m, M])$.
Using (4.40) and properties (P) and (ii), we derive the operator Poincaré inequality:

Theorem 4.17 *All as in Theorem 4.16. Then*

$$\int_{m1_H}^{A} \left|D_{m}^{\gamma}f\right|^{q} \leq$$
$$\left[\frac{(A - m1_H)^{q(\nu-\gamma)}}{(\Gamma(\nu-\gamma))^{q}(p(\nu-\gamma-1)+1)^{\frac{q}{p}} q(\nu-\gamma)}\right]\left(\int_{m1_H}^{A}\left|D_{m}^{\nu}f\right|^{q}\right). \tag{4.41}$$

We will use

Theorem 4.18 ([1], p. 485) *Let $\nu \geq \gamma + 1$, $\gamma \geq 0$, $n := [\nu]$. Assume $f \in C_{m}^{\nu}([m, M])$ such that $f^{(k)}(m) = 0$, $k = 0, 1, \ldots, n-1$. Let $p, q > 1 : \frac{1}{p} + \frac{1}{q} = 1$, $r \geq 1$. Then*

$$\left(\int_{m}^{\lambda}\left|D_{m}^{\gamma}f(t)\right|^{r} dt\right)^{\frac{1}{r}} \leq$$
$$\left[\frac{(\lambda - m)^{\nu-\gamma+\frac{1}{r}-\frac{1}{q}}}{(\Gamma(\nu-\gamma))(p(\nu-\gamma-1)+1)^{\frac{1}{p}}}\right]\frac{\left(\int_{m}^{\lambda}\left|D_{m}^{\nu}f(t)\right|^{q} dt\right)^{\frac{1}{q}}}{\left[r\left(\nu-\gamma-\frac{1}{q}\right)+1\right]^{\frac{1}{r}}}, \tag{4.42}$$

$\forall \lambda \in [m, M]$.

Applying (4.42), using properties (P) and (ii), we get the following operator Sobolev type inequality:

Theorem 4.19 *All as in Theorem 4.18. Then*

$$\left(\int_{m1_H}^{A}\left|D_{m}^{\gamma}f\right|^{r}\right)^{\frac{1}{r}} \leq$$
$$\frac{(A - m1_H)^{\nu-\gamma+\frac{1}{r}-\frac{1}{q}}}{(\Gamma(\nu-\gamma))(p(\nu-\gamma-1)+1)^{\frac{1}{p}}}\frac{\left(\int_{m1_H}^{A}\left|D_{m}^{\nu}f\right|^{q}\right)^{\frac{1}{q}}}{\left[r\left(\nu-\gamma-\frac{1}{q}\right)+1\right]^{\frac{1}{r}}}. \tag{4.43}$$

We need

Definition 4.20 ([2], *p. 337*) Let $f \in AC^{n}([m, M])$, $n := \lceil \nu \rceil$, $\nu > 0$. The right Caputo fractional derivative of order $\nu > 0$, is given by

$$\left(D_{M-}^{\nu}f\right)(z) := \frac{(-1)^{n}}{\Gamma(n-\nu)}\int_{z}^{M}(J - z)^{n-\nu-1} f^{(n)}(J)\, dJ, \tag{4.44}$$

$\forall z \in [m, M]$, which exists a.e. on $[m, M]$, and $D^{\nu}_{M-} f \in L_1([m, M])$.

We notice that $D^0_{M-} f = f$, $\left(D^n_{M-} f\right)(z) = (-1)^n f^{(n)}(z)$, for $n \in \mathbb{N}$.

We need the right Caputo fractional Taylor formula with integral remainder:

Theorem 4.21 ([2], p. 341) *Let $f \in AC^n([m, M])$, $\lambda \in [m, M]$, $\nu > 0$, $n = \lceil \nu \rceil$. Then*

$$f(\lambda) = \sum_{k=0}^{n-1} \frac{f^{(k)}(M)}{k!} (\lambda - M)^k + \frac{1}{\Gamma(\nu)} \int_{\lambda}^{M} (J - \lambda)^{\nu-1} \left(D^{\nu}_{M-} f\right)(J)\, dJ. \tag{4.45}$$

We present the following operator representation formula:

Theorem 4.22 *Let A be a selfadjoint operator in the Hilbert space H with the spectrum $Sp(A) \subseteq [m, M]$ for some real numbers $m < M$, $\{E_{\lambda}\}_{\lambda}$ be its spectral family, I be a closed subinterval on $\mathbb{R}$ with $[m, M] \subset \overset{\circ}{I}$ (the interior of I) and $n \in \mathbb{N}$, with $n := \lceil \nu \rceil$, $\nu > 0$. We consider $f \in AC^n([m, M])$ (i.e. $f^{(n-1)} \in AC([m, M])$), where $f : I \to \mathbb{R}$.*

Then

$$f(A) = \sum_{k=0}^{n-1} \frac{f^{(k)}(M)}{k!} (A - M 1_H)^k +$$
$$\frac{1}{\Gamma(\nu)} \int_{m-0}^{M} \left(\int_{\lambda}^{M} (J - \lambda)^{\nu-1} \left(D^{\nu}_{M-} f\right)(J)\, dJ \right) dE_{\lambda}. \tag{4.46}$$

Proof Integrate (4.45) against E_{λ} and apply the spectral representation theorem. ■

We make

Remark 4.23 In (4.46) assume that $f^{(k)}(M) = 0$, $k = 0, \ldots, n-1$. Then

$$f(A) = \frac{1}{\Gamma(\nu)} \int_{m-0}^{M} \left(\int_{\lambda}^{M} (J - \lambda)^{\nu-1} \left(D^{\nu}_{M-} f\right)(J)\, dJ \right) dE_{\lambda}. \tag{4.47}$$

We have that

$$\langle f(A)x, x \rangle = \frac{1}{\Gamma(\nu)} \int_{m-0}^{M} \left(\int_{\lambda}^{M} (J - \lambda)^{\nu-1} \left(D^{\nu}_{M-} f\right)(J)\, dJ \right) d\langle E_{\lambda}x, x \rangle, \tag{4.48}$$

$\forall x \in H$.

Let $p, q > 1 : \frac{1}{p} + \frac{1}{q} = 1$, with $\nu > \frac{1}{q}$. Then

$$\int_{\lambda}^{M} (J-\lambda)^{\nu-1} \left|\left(D_{M-}^{\nu} f\right)(J)\right| dJ \leq$$

$$\left(\int_{\lambda}^{M} (J-\lambda)^{p(\nu-1)} dJ\right)^{\frac{1}{p}} \left(\int_{\lambda}^{M} \left|\left(D_{M-}^{\nu} f\right)(J)\right|^{q} dJ\right)^{\frac{1}{q}} \leq \tag{4.49}$$

$$\frac{(M-\lambda)^{\frac{p(\nu-1)+1}{p}}}{(p(\nu-1)+1)^{\frac{1}{p}}} \left\|D_{M-}^{\nu} f\right\|_{q,[m,M]} = \frac{(M-\lambda)^{\nu-\frac{1}{q}}}{(p(\nu-1)+1)^{\frac{1}{p}}} \left\|D_{M-}^{\nu} f\right\|_{q,[m,M]}.$$

We have proved that

$$\left|\int_{\lambda}^{M} (J-\lambda)^{\nu-1} \left(D_{M-}^{\nu} f\right)(J) dJ\right| \leq \int_{\lambda}^{M} (J-\lambda)^{\nu-1} \left|\left(D_{M-}^{\nu} f\right)(J)\right| dJ \leq$$

$$\frac{(M-\lambda)^{\nu-\frac{1}{q}}}{(p(\nu-1)+1)^{\frac{1}{p}}} \left\|D_{M-}^{\nu} f\right\|_{q,[m,M]}, \tag{4.50}$$

$\forall \lambda \in [m, M]$.

Therefore it holds

$$|\langle f(A)x, x\rangle| \overset{(4.48)}{\leq} \frac{1}{\Gamma(\nu)} \int_{m-0}^{M} \left(\int_{\lambda}^{M} (J-\lambda)^{\nu-1} \left(D_{M-}^{\nu} f\right)(J) dJ\right) d\langle E_{\lambda}x, x\rangle$$

$$\overset{(4.50)}{\leq} \frac{\left\|D_{M-}^{\nu} f\right\|_{q,[m,M]}}{\Gamma(\nu)(p(\nu-1)+1)^{\frac{1}{p}}} \int_{m-0}^{M} (M-\lambda)^{\nu-\frac{1}{q}} d\langle E_{\lambda}x, x\rangle = \tag{4.51}$$

$$\frac{\left\|D_{M-}^{\nu} f\right\|_{q,[m,M]}}{(p(\nu-1)+1)^{\frac{1}{p}} \Gamma(\nu)} \left\langle (M1_H - A)^{\nu-\frac{1}{q}} x, x\right\rangle, \quad \forall x \in H.$$

We have proved

Theorem 4.24 *All as in Theorem 4.22. Assume further $f^{(k)}(M) = 0$, $k = 0, 1, \ldots, n-1$. Let $p, q > 1 : \frac{1}{p} + \frac{1}{q} = 1$, with $\nu > \frac{1}{q}$. Then*

$$|\langle f(A)x, x\rangle| \leq \frac{\left\|D_{M-}^{\nu} f\right\|_{q,[m,M]}}{(p(\nu-1)+1)^{\frac{1}{p}} \Gamma(\nu)} \left\langle (M1_H - A)^{\nu-\frac{1}{q}} x, x\right\rangle, \tag{4.52}$$

$\forall x \in H$.

Inequality (4.52) means

$$\|f(A)\| \leq \frac{\left\|D_{M-}^{\nu} f\right\|_{q,[m,M]}}{(p(\nu-1)+1)^{\frac{1}{p}} \Gamma(\nu)} \left\|(M1_H - A)^{\nu-\frac{1}{q}}\right\|, \tag{4.53}$$

and in particular,

$$f(A) \leq \left(\frac{\left\| D_{M-}^{\nu} f \right\|_{q,[m,M]}}{(p(\nu-1)+1)^{\frac{1}{p}} \Gamma(\nu)} \right) (M1_H - A)^{\nu - \frac{1}{q}}. \tag{4.54}$$

We give the following Poincaré type fractional inequality:

Theorem 4.25 *Let $f \in AC^n([m,M])$, $\nu > 0$, $n = \lceil \nu \rceil$. Assume $f^{(k)}(M) = 0$, $k = 0, \ldots, n-1$. Let $p, q > 1 : \frac{1}{p} + \frac{1}{q} = 1$, $\nu > \frac{1}{q}$. Then*

$$\int_w^M |f(\lambda)|^q d\lambda \leq \frac{(M-w)^{\nu q}}{(p(\nu-1)+1)^{\frac{q}{p}} (\Gamma(\nu))^q \nu q} \int_w^M \left| \left(D_{M-}^{\nu} f \right)(\lambda) \right|^q d\lambda, \tag{4.55}$$

$\forall\, w \in [m,M]$.

Proof By the assumption and (4.45) we have that

$$f(\lambda) = \frac{1}{\Gamma(\nu)} \int_\lambda^M (J-\lambda)^{\nu-1} \left(D_{M-}^{\nu} f \right)(J)\, dJ, \ \forall\, \lambda \in [m,M]. \tag{4.56}$$

Hence

$$|f(\lambda)| \leq \frac{1}{\Gamma(\nu)} \int_\lambda^M (J-\lambda)^{\nu-1} \left| \left(D_{M-}^{\nu} f \right)(J) \right| dJ, \ \forall\, \lambda \in [m,M]. \tag{4.57}$$

As in (4.49), (4.50), we get

$$|f(\lambda)| \leq \frac{(M-\lambda)^{\nu-\frac{1}{q}}}{(p(\nu-1)+1)^{\frac{1}{p}} \Gamma(\nu)} \left\| D_{M-}^{\nu} f \right\|_{q,[w,M]}, \tag{4.58}$$

$\forall\, \lambda \in [w,M]$, where $w \in [m,M]$.

Hence it holds

$$|f(\lambda)|^q \leq \frac{(M-\lambda)^{\nu q-1}}{(p(\nu-1)+1)^{\frac{q}{p}} (\Gamma(\nu))^q} \left\| D_{M-}^{\nu} f \right\|_{q,[w,M]}^q, \tag{4.59}$$

$\forall\, \lambda \in [w,M]$, where $w \in [m,M]$.

Therefore by integration

$$\int_w^M |f(\lambda)|^q d\lambda \leq \frac{(M-w)^{\nu q}}{(p(\nu-1)+1)^{\frac{q}{p}} (\Gamma(\nu))^q \nu q} \left\| D_{M-}^{\nu} f \right\|_{q,[w,M]}^q, \tag{4.60}$$

$\forall\, w \in [m,M]$, proving the claim. ■

We need

Definition 4.26 Let $f : [m,M] \to \mathbb{R}$ be continuous. We consider

$$g(t) = \int_t^M f(z)\,dz, \forall\, t \in [m, M], \tag{4.61}$$

then $g \in C([m, M])$.

We denote by

$$\int_A^{M1_H} f := \Phi(g) = g(A). \tag{4.62}$$

We denote also

$$g^r(A) = \Phi(g^r) =: \left(\int_A^{M1_H} f\right)^r, \quad r > 0. \tag{4.63}$$

Clearly $\left(\int_A^{M1_H} f\right)^r$ is a self adjoint operator on H, for any $r > 0$.

We present the following operator Poincaré type inequality:

Theorem 4.27 *All as in Theorem 4.25. Then*

$$\int_A^{M1_H} |f|^q \leq \frac{(M1_H - A)^{\nu q}}{(p(\nu-1)+1)^{\frac{q}{p}} (\Gamma(\nu))^q \nu q} \left(\int_A^{M1_H} \left|\left(D_{M-}^{\nu} f\right)\right|^q\right). \tag{4.64}$$

We give the following Sobolev type fractional inequality:

Theorem 4.28 *All as in Theorem 4.25, and $r \geq 1$. Then*

$$\|f\|_{r,[w,M]} \leq \frac{(M-w)^{\nu - \frac{1}{q} + \frac{1}{r}}}{\left(\nu r - \frac{r}{p} + 1\right)^{\frac{1}{r}} (p(\nu-1)+1)^{\frac{1}{p}} \Gamma(\nu)} \left\|D_{M-}^{\nu} f\right\|_{q,[w,M]}, \tag{4.65}$$

$\forall\, w \in [m, M]$.

Proof We recall (4.58):

$$|f(\lambda)| \leq \frac{(M-\lambda)^{\nu - \frac{1}{q}}}{\Gamma(\nu)(p(\nu-1)+1)^{\frac{1}{p}}} \left\|D_{M-}^{\nu} f\right\|_{q,[w,M]}, \tag{4.66}$$

$\forall\, \lambda \in [w, M]$, where $w \in [m, M]$.

Hence, by $r \geq 1$, we obtain

$$|f(\lambda)|^r \leq \frac{(M-\lambda)^{\nu r - \frac{r}{q}}}{(\Gamma(\nu))^r (p(\nu-1)+1)^{\frac{r}{p}}} \left\|D_{M-}^{\nu} f\right\|_{q,[w,M]}^r, \tag{4.67}$$

$\forall\, \lambda \in [w, M]$, where $w \in [m, M]$.

Consequently it holds

$$\int_{w}^{M} |f(\lambda)|^r d\lambda \leq \frac{(M-w)^{\nu r - \frac{r}{q} + 1}}{\left(\nu r - \frac{r}{q} + 1\right) (p(\nu-1)+1)^{\frac{r}{p}} (\Gamma(\nu))^r} \left\| D_{M-}^{\nu} f \right\|_{q,[w,M]}^{r}, \tag{4.68}$$

$\forall\, w \in [m, M]$, proving the claim. ■

Next we give an operator Sobolev type inequality:

Theorem 4.29 *All as in Theorem 4.28. Then*

$$\left(\int_{A}^{M1_H} |f|^r \right)^{\frac{1}{r}} \leq \frac{(M1_H - A)^{\nu - \frac{1}{q} + \frac{1}{r}}}{\left(\nu r - \frac{r}{p} + 1\right)^{\frac{1}{r}} (p(\nu-1)+1)^{\frac{1}{p}} \Gamma(\nu)} \left(\int_{A}^{M1_H} \left|\left(D_{M-}^{\nu} f\right)\right|^q \right)^{\frac{1}{q}}. \tag{4.69}$$

We need

Definition 4.30 ([2], *p. 345*) Let $\nu > 0$, $n := [\nu]$, $\alpha := \nu - n$, $0 < \alpha < 1$, $f \in C([m, M])$. The right Riemann–Liouville fractional integral operator in given by

$$\left(J_{M-}^{\nu} f\right)(z) := \frac{1}{\Gamma(\nu)} \int_{z}^{M} (J-z)^{\nu-1} f(J)\, dJ, \tag{4.70}$$

$\forall\, z \in [m, M]$, $J_{M-}^{0} f := f$.

Define the subspace of functions

$$C_{M-}^{\nu}([m, M]) := \left\{ f \in C^n([m, M]) : J_{M-}^{1-\alpha} f^{(n)} \in C^1([m, M]) \right\}. \tag{4.71}$$

Define the right generalized ν-fractional derivative of f over $[m, M]$ as

$$\overline{D}_{M-}^{\nu} f := (-1)^{n-1} \left(J_{M-}^{1-\alpha} f^{(n)}\right)'. \tag{4.72}$$

Notice that

$$J_{M-}^{1-\alpha} f^{(n)}(z) = \frac{1}{\Gamma(1-\alpha)} \int_{z}^{M} (J-z)^{-\alpha} f^{(n)}(J)\, dJ, \tag{4.73}$$

exists for $f \in C_{M-}^{\nu}([m, M])$, and

$$\left(\overline{D}_{M-}^{\nu} f\right)(z) = \frac{(-1)^{n-1}}{\Gamma(1-\alpha)} \frac{d}{dz} \int_{z}^{M} (J-z)^{-\alpha} f^{(n)}(J)\, dJ. \tag{4.74}$$

That is

$$\left(\overline{D}_{M-}^{\nu} f\right)(z) = \frac{(-1)^{n-1}}{\Gamma(n-\nu+1)} \frac{d}{dz} \int_{z}^{M} (J-z)^{n-\nu} f^{(n)}(J)\, dJ. \tag{4.75}$$

If $\nu \in \mathbb{N}$, then $\alpha = 0$, $n = \nu$, and

$$\left(\overline{D}_{M-}^{\nu} f\right)(z) = (-1)^n f^{(n)}(z), \tag{4.76}$$

$\forall\, z \in [m, M]$, and $\overline{D}_{M-}^{0} f = f$.

We will use the following fractional Taylor formula:

Theorem 4.31 ([2], p. 348) *Let $f \in C_{M-}^{\nu}([m, M])$, $\nu > 0$, $n := [\nu]$. Then*
(1) for $\nu \geq 1$, we get

$$f(\lambda) = \sum_{k=0}^{n-1} \frac{f^{(k)}(M)}{k!} (\lambda - M)^k + \frac{1}{\Gamma(\nu)} \int_{\lambda}^{M} (J - \lambda)^{\nu-1} \left(\overline{D}_{M-}^{\nu} f\right)(J)\, dJ, \tag{4.77}$$

$\forall\, \lambda \in [m, M]$.
(2) if $0 < \nu < 1$, we obtain

$$f(\lambda) = \frac{1}{\Gamma(\nu)} \int_{\lambda}^{M} (J - \lambda)^{\nu-1} \left(\overline{D}_{M-}^{\nu} f\right)(J)\, dJ, \tag{4.78}$$

$\forall\, \lambda \in [m, M]$.

We present the following operator representation formula:

Theorem 4.32 *Let A be a selfadjoint operator in the Hilbert space H with the spectrum $Sp(A) \subseteq [m, M]$ for some real numbers $m < M$, $\{E_\lambda\}_\lambda$ be its spectral family, I be a closed subinterval on $\mathbb{R}$ with $[m, M] \subset \overset{\circ}{I}$ (the interior of I) and $n \in \mathbb{N}$, with $n := [\nu]$, $\nu > 0$. We consider $f \in C_{M-}^{\nu}([m, M])$, where $f : I \to \mathbb{R}$. Then*

(i) case of $\nu \geq 1$,

$$f(A) = \sum_{k=0}^{n-1} \frac{f^{(k)}(M)}{k!} (A - M 1_H)^k +$$
$$\frac{1}{\Gamma(\nu)} \int_{m-0}^{M} \left(\int_{\lambda}^{M} (J - \lambda)^{\nu-1} \left(\overline{D}_{M-}^{\nu} f\right)(J)\, dJ \right) dE_\lambda, \tag{4.79}$$

(ii) case of $0 < \nu < 1$,

$$f(A) = \frac{1}{\Gamma(\nu)} \int_{m-0}^{M} \left(\int_{\lambda}^{M} (J - \lambda)^{\nu-1} \left(\overline{D}_{M-}^{\nu} f\right)(J)\, dJ \right) dE_\lambda. \tag{4.80}$$

Proof Integrate (4.77), (4.78) against E_λ, apply spectral representation theorem. ■

We have proved

Theorem 4.33 *All as in Theorem 4.32. In case of $\nu \geq 1$, we assume further $f^{(k)}(M) = 0$, for $k = 0, 1, \ldots, n-1$. Let $p, q > 1 : \frac{1}{p} + \frac{1}{q} = 1$, with $\nu > \frac{1}{q}$. Then*

$$|\langle f(A)x, x\rangle| \leq \frac{\left\| \overline{D}_{M-}^{\nu} f \right\|_{q,[m,M]}}{(p(\nu-1)+1)^{\frac{1}{p}} \Gamma(\nu)} \left\langle (M1_H - A)^{\nu - \frac{1}{q}} x, x \right\rangle, \tag{4.81}$$

$\forall x \in H$.

Inequality (4.81) means

$$\|f(A)\| \leq \frac{\left\| \overline{D}_{M-}^{\nu} f \right\|_{q,[m,M]}}{(p(\nu-1)+1)^{\frac{1}{p}} \Gamma(\nu)} \left\| (M1_H - A)^{\nu - \frac{1}{q}} \right\|, \tag{4.82}$$

and in particular,

$$f(A) \leq \left(\frac{\left\| \overline{D}_{M-}^{\nu} f \right\|_{q,[m,M]}}{(p(\nu-1)+1)^{\frac{1}{p}} \Gamma(\nu)} \right) (M1_H - A)^{\nu - \frac{1}{q}}. \tag{4.83}$$

Proof Very similar to Theorem 4.24. ■

We give the following Poincaré type fractional inequality:

Theorem 4.34 *Let $f \in C_{M-}^{\nu}([m, M])$, $\nu > 0$, $n = [\nu]$. If $\nu \geq 1$, we assume $f^{(k)}(M) = 0$, $k = 0, \ldots, n-1$. Let $p, q > 1 : \frac{1}{p} + \frac{1}{q} = 1$, $\nu > \frac{1}{q}$. Then*

$$\int_w^M |f(\lambda)|^q \, d\lambda \leq \frac{(M-w)^{\nu q}}{(p(\nu-1)+1)^{\frac{q}{p}} (\Gamma(\nu))^q \nu q} \int_w^M \left| \left(\overline{D}_{M-}^{\nu} f \right)(\lambda) \right|^q d\lambda, \tag{4.84}$$

$\forall \, w \in [m, M]$.

Proof Similar to Theorem 4.25. ■

We present the following operator Poincaré type inequality:

Theorem 4.35 *All as in Theorem 4.34. Then*

$$\int_A^{M1_H} |f|^q \leq \frac{(M1_H - A)^{\nu q}}{(p(\nu-1)+1)^{\frac{q}{p}} (\Gamma(\nu))^q \nu q} \left(\int_A^{M1_H} \left| \overline{D}_{M-}^{\nu} f \right|^q \right). \tag{4.85}$$

We give the following Sobolev type fractional inequality:

Theorem 4.36 *All as in Theorem 4.34, and $r \geq 1$. Then*

$$\|f\|_{r,[w,M]} \leq \frac{(M-w)^{\nu-\frac{1}{q}+\frac{1}{r}}}{\left(\nu r-\frac{r}{p}+1\right)^{\frac{1}{r}}(p(\nu-1)+1)^{\frac{1}{p}}\Gamma(\nu)}\left\|\overline{D}_{M-}^{\nu}f\right\|_{q,[w,M]}, \quad (4.86)$$

$\forall\, w \in [m, M]$.

Proof Similar to Theorem 4.28. ∎

Next we give an operator Sobolev type inequality:

Theorem 4.37 *All as in Theorem 4.36. Then*

$$\left(\int_A^{M1_H}|f|^r\right)^{\frac{1}{r}} \leq \frac{(M1_H-A)^{\nu-\frac{1}{q}+\frac{1}{r}}}{\left(\nu r-\frac{r}{p}+1\right)^{\frac{1}{r}}(p(\nu-1)+1)^{\frac{1}{p}}\Gamma(\nu)}\left(\int_A^{M1_H}\left|\overline{D}_{M-}^{\nu}f\right|^q\right)^{\frac{1}{q}}. \quad (4.87)$$

References

1. G.A. Anastassiou, *Fractional Differentiation Inequalities* (Springer, New York, 2009)
2. G.A. Anastassiou, *Intelligent Mathematics: Computational Analysis* (Springer, Heidelberg, 2011)
3. G. Anastassiou, *Fractional Self adjoint Operator Poincaré and Sobolev type Inequalities*, Fasciculi Mathematici (2016, accepted)
4. K. Diethelm, *The Analysis of Fractional Differential Equations* (Springer, New York, 2010)
5. S.S. Dragomir, *Inequalities for functions of selfadjoint operators on Hilbert Spaces* (2011), ajmaa.org/RGMIA/monographs/InFuncOp.pdf
6. S. Dragomir, *Operator Inequalities of Ostrowski and Trapezoidal type* (Springer, New York, 2012)
7. T. Furuta, J. Mićić Hot, J. Pečaric, Y. Seo, *Mond-Pečaric Method in Operator Inequalities. Inequalities for Bounded Selfadjoint Operators on a Hilbert Space* (Element, Zagreb, 2005)
8. G. Helmberg, *Introduction to Spectral Thery in Hilbert Space* (Wiley, New York, 1969)

Chapter 5
Self Adjoint Operator Ostrowski Inequalities

We present here several self adjoint operator Ostrowski type inequalities to all directions. These are based in the operator order over a Hilbert space. It follows [5].

5.1 Motivation

In 1938, A. Ostrowski [13] proved the following important inequality:

Let $f : [a, b] \to \mathbb{R}$ be continuous on $[a, b]$ and differentiable on (a, b) whose derivative $f' : (a, b) \to \mathbb{R}$ is bounded on (a, b), i.e., $\left\| f' \right\|_{\infty} := \sup_{t \in (a,b)} \left| f'(t) \right| < +\infty$. Then

$$\left| \frac{1}{b-a} \int_a^b f(t)\, dt - f(x) \right| \leq \left[\frac{1}{4} + \frac{\left(x - \frac{a+b}{2} \right)^2}{(b-a)^2} \right] (b-a) \left\| f' \right\|_{\infty},$$

for any $x \in [a, b]$. The constant $\frac{1}{4}$ is the best possible.

In this chapter we present self adjoint operator Ostrowski type inequalities on a Hilbert space in the operator order.

5.2 Background

Let A be a selfadjoint linear operator on a complex Hilbert space $(H; \langle \cdot, \cdot \rangle)$. The Gelfand map establishes a $*-$isometrically isomorphism Φ between the set $C(Sp(A))$ of all continuous functions defined on the spectrum of A, denoted $Sp(A)$, and the C^*-algebra $C^*(A)$ generated by A and the identity operator 1_H on H as follows (see e.g. [11, p. 3]):

G.A. Anastassiou, *Intelligent Comparisons II: Operator Inequalities and Approximations*, Studies in Computational Intelligence 699,
DOI 10.1007/978-3-319-51475-8_5

For any $f, g \in C(Sp(A))$ and any $\alpha, \beta \in \mathbb{C}$ we have

(i) $\Phi(\alpha f + \beta g) = \alpha\Phi(f) + \beta\Phi(g)$;
(ii) $\Phi(fg) = \Phi(f)\Phi(g)$ (the operation composition is on the right) and $\Phi(\overline{f}) = (\Phi(f))^*$;
(iii) $\|\Phi(f)\| = \|f\| := \sup_{t \in Sp(A)} |f(t)|$;
(iv) $\Phi(f_0) = 1_H$ and $\Phi(f_1) = A$, where $f_0(t) = 1$ and $f_1(t) = t$, for $t \in Sp(A)$.

With this notation we define

$$f(A) := \Phi(f), \text{ for all } f \in C(Sp(A)),$$

and we call it the continuous functional calculus for a selfadjoint operator A.

If A is a selfadjoint operator and f is a real valued continuous function on $Sp(A)$ then $f(t) \geq 0$ for any $t \in Sp(A)$ implies that $f(A) \geq 0$, i.e. $f(A)$ is a positive operator on H. Moreover, if both f and g are real valued continuous functions on $Sp(A)$ then the following important property holds:

(P) $f(t) \geq g(t)$ for any $t \in Sp(A)$, implies that $f(A) \geq g(A)$ in the operator order of $B(H)$ (the Banach algebra of all bounded linear operators from H into itself).

Equivalently, we use (see [9], pp. 7–8):

Let U be a selfadjoint operator on the complex Hilbert space $(H, \langle\cdot,\cdot\rangle)$ with the spectrum $Sp(U)$ included in the interval $[m, M]$ for some real numbers $m < M$ and $\{E_\lambda\}_\lambda$ be its spectral family.

Then for any continuous function $f : [m, M] \to \mathbb{C}$, it is well known that we have the following spectral representation in terms of the Riemann–Stieljes integral:

$$\langle f(U)x, y\rangle = \int_{m-0}^{M} f(\lambda)\, d(\langle E_\lambda x, y\rangle),$$

for any $x, y \in H$. The function $g_{x,y}(\lambda) := \langle E_\lambda x, y\rangle$ is of bounded variation on the interval $[m, M]$, and

$$g_{x,y}(m-0) = 0 \text{ and } g_{x,y}(M) = \langle x, y\rangle,$$

for any $x, y \in H$. Furthermore, it is known that $g_x(\lambda) := \langle E_\lambda x, x\rangle$ is increasing and right continuous on $[m, M]$.

We have also the formula

$$\langle f(U)x, x\rangle = \int_{m-0}^{M} f(\lambda)\, d(\langle E_\lambda x, x\rangle), \ \forall\, x \in H.$$

As a symbol we can write

$$f(U) = \int_{m-0}^{M} f(\lambda)\, dE_{\lambda}.$$

Above, $m = \min\{\lambda | \lambda \in Sp(U)\} := \min Sp(U)$, $M = \max\{\lambda | \lambda \in Sp(U)\} := \max Sp(U)$. The projections $\{E_{\lambda}\}_{\lambda \in \mathbb{R}}$, are called the spectral family of A, with the properties:

(a) $E_{\lambda} \leq E_{\lambda'}$ for $\lambda \leq \lambda'$;
(b) $E_{m-0} = 0_H$ (zero operator), $E_M = 1_H$ (identity operator) and $E_{\lambda+0} = E_{\lambda}$ for all $\lambda \in \mathbb{R}$.

Furthermore

$$E_{\lambda} := \varphi_{\lambda}(U), \forall\, \lambda \in \mathbb{R},$$

is a projection which reduces U, with

$$\varphi_{\lambda}(s) := \begin{cases} 1, \text{ for } -\infty < s \leq \lambda, \\ 0, \text{ for } \lambda < s < +\infty. \end{cases}$$

The spectral family $\{E_{\lambda}\}_{\lambda \in \mathbb{R}}$ determines uniquely the self-adjoint operator U and vice versa.

For more on the topic see [12], pp. 256–266, and for more details see there pp. 157–266. See also [8].

Some more basics are given (we follow [9], pp. 1–5):

Let $(H; \langle \cdot, \cdot \rangle)$ be a Hilbert space over $\mathbb{C}$. A bounded linear operator A defined on H is selfjoint, i.e., $A = A^*$, iff $\langle Ax, x \rangle \in \mathbb{R}$, $\forall\, x \in H$, and if A is selfadjoint, then

$$\|A\| = \sup_{x \in H : \|x\| = 1} |\langle Ax, x \rangle|.$$

Let A, B be selfadjoint operators on H. Then $A \leq B$ iff $\langle Ax, x \rangle \leq \langle Bx, x \rangle, \forall\, x \in H$.

In particular, A is called positive if $A \geq 0$.

Denote by

$$\mathcal{P} := \left\{ \varphi(s) := \sum_{k=0}^{n} \alpha_k s^k | n \geq 0, \alpha_k \in \mathbb{C}, 0 \leq k \leq n \right\}.$$

If $A \in \mathcal{B}(H)$ is selfadjoint, and $\varphi(s) \in \mathcal{P}$ has real coefficients, then $\varphi(A)$ is selfadjoint, and

$$\|\varphi(A)\| = \max\{|\varphi(\lambda)|, \lambda \in Sp(A)\}.$$

If φ is any function defined on $\mathbb{R}$ we define

$$\|\varphi\|_A := \sup\{|\varphi(\lambda)|, \lambda \in Sp(A)\}.$$

If A is selfadjoint operator on Hilbert space H and φ is continuous and given that $\varphi(A)$ is selfadjoint, then $\|\varphi(A)\| = \|\varphi\|_A$. And if φ is a continuous real valued function so it is $|\varphi|$, then $\varphi(A)$ and $|\varphi|(A) = |\varphi(A)|$ are selfadjoint operators (by [9], p. 4, Theorem 7).

Hence it holds

$$\begin{aligned} \||\varphi(A)|\| &= \||\varphi|\|_A = \sup\{||\varphi(\lambda)||, \lambda \in Sp(A)\} \\ &= \sup\{|\varphi(\lambda)|, \lambda \in Sp(A)\} = \|\varphi\|_A = \|\varphi(A)\|, \end{aligned}$$

that is

$$\||\varphi(A)|\| = \|\varphi(A)\|.$$

For a selfadjoint operator $A \in \mathcal{B}(H)$ which is positive, there exists a unique positive selfadjoint operator $B := \sqrt{A} \in \mathcal{B}(H)$ such that $B^2 = A$, that is $\left(\sqrt{A}\right)^2 = A$. We call B the square root of A.

Let $A \in \mathcal{B}(H)$, then A^*A is selfadjoint and positive. Define the "operator absolute value" $|A| := \sqrt{A^*A}$. If $A = A^*$, then $|A| = \sqrt{A^2}$.

For a continuous real valued function φ we observe the following:

$$|\varphi(A)| \text{ (the functional absolute value) } = \int_{m-0}^{M} |\varphi(\lambda)|\, dE_\lambda =$$

$$\int_{m-0}^{M} \sqrt{(\varphi(\lambda))^2} dE_\lambda = \sqrt{(\varphi(A))^2} = |\varphi(A)| \text{ (operator absolute value)},$$

where A is a selfadjoint operator.

That is we have

$$|\varphi(A)| \text{ (functional absolute value) } = |\varphi(A)| \text{ (operator absolute value)}.$$

5.3 Main Results

Let A be a selfadjoint operator in the Hilbert space H with the spectrum $Sp(A) \subseteq [m, M]$, $m < M$; $m, M \in \mathbb{R}$.

In the next we obtain Ostrowski type inequalities in the operator order of $\mathcal{B}(H)$ (the Banach algebra of all bounded linear operators from H into itself).

We mention

Theorem 5.1 ([2], p. 498) *Let $f \in C^1([m, M])$, $m < M$, $s \in [m, M]$. Then*

$$\left| \frac{1}{M-m} \int_m^M f(t)\,dt - f(x) \right| \le \left(\frac{(s-m)^2 + (M-s)^2}{2(M-m)} \right) \|f'\|_\infty . \tag{5.1}$$

By applying property (P) to (5.1), we obtain in the operator order the following inequality:

Theorem 5.2 *Let $f \in C^1([m, M])$. Then*

$$\left| \left(\frac{1}{M-m} \int_m^M f(t)\,dt \right) 1_H - f(A) \right| \le \left(\frac{(A - m1_H)^2 + (M1_H - A)^2}{2(M-m)} \right) \|f'\|_\infty . \tag{5.2}$$

We mention

Theorem 5.3 ([1], p. 191, Cerone–Dragomir) *Let $f : [m, M] \to \mathbb{R}$ be a continuous on $[m, M]$ and twice differentiable function on (m, M), whose second derivative $f'' : (m, M) \to \mathbb{R}$ is bounded on (m, M). Then*

$$\left| f(s) - \frac{1}{M-m} \int_m^M f(t)\,dt - \left(\frac{f(M) - f(m)}{M-m} \right) \left(s - \frac{m+M}{2} \right) \right| \le \tag{5.3}$$

$$\frac{1}{2} \left\{ \left[\frac{\left(s - \frac{m+M}{2}\right)^2}{(M-m)^2} + \frac{1}{4} \right]^2 + \frac{1}{12} \right\} (M-m)^2 \|f''\|_\infty \le \frac{\|f''\|_\infty}{6} (M-m)^2,$$

$\forall\, s \in [m, M]$.

By applying property (P) to (5.3), we obtain in the operator order the following inequality:

Theorem 5.4 *All as in Theorem 5.3. Then*

$$\left| f(A) - \left(\frac{1}{M-m} \int_m^M f(t)\,dt \right) 1_H - \left(\frac{f(M) - f(m)}{M-m} \right) \left(A - \left(\frac{m+M}{2} \right) 1_H \right) \right| \tag{5.4}$$

$$\le \frac{1}{2} \left\{ \left[\frac{\left(A - \left(\frac{m+M}{2}\right) 1_H\right)^2}{(M-m)^2} + \frac{1}{4} 1_H \right]^2 + \frac{1}{12} 1_H \right\} (M-m)^2 \|f''\|_\infty$$

$$\le \left(\frac{\|f''\|_\infty}{6} (M-m)^2 \right) 1_H .$$

We mention

Theorem 5.5 ([3], p. 14) *Let $f : [m, M] \to \mathbb{R}$ be 3-times differentiable on $[m, M]$. Assume that f''' is bounded on $[m, M]$. Let any $s \in [m, M]$. Then*

$$\left| f(s) - \frac{1}{M-m} \int_m^M f(t)\,dt - \left(\frac{f(M) - f(m)}{M-m} \right) \left(s - \left(\frac{m+M}{2} \right) \right) - \right.$$
$$\left. \left(\frac{f'(M) - f'(m)}{2(M-m)} \right) \left[s^2 - (m+M)s + \left(\frac{m^2 + M^2 + 4mM}{6} \right) \right] \right| \tag{5.5}$$
$$\leq \frac{\|f'''\|_\infty}{(M-m)^3} Z(s),$$

where

$$Z(s) = \left[mMs^4 - \frac{1}{3}m^2M^3s + \frac{1}{3}m^3Ms^2 - mM^2s^3 - \frac{1}{3}m^3M^2s + \frac{1}{3}mM^3s^2 \right.$$
$$+ m^2M^2s^2 - m^2Ms^3 - \frac{1}{2}ms^5 - \frac{1}{2}Ms^5 + \frac{1}{6}s^6 + \frac{3}{4}m^2s^4 + \frac{3}{4}M^2s^4 + \frac{1}{3}M^2m^4 -$$
$$\frac{2}{3}m^3s^3 - \frac{2}{3}M^3s^3 - \frac{1}{3}M^3m^3 + \frac{5}{12}m^4s^2 + \frac{5}{12}M^4s^2 + \frac{1}{3}M^4m^2 -$$
$$\left. \frac{2}{15}Mm^5 - \frac{2}{15}mM^5 - \frac{1}{6}m^5s - \frac{1}{6}M^5s + \frac{m^6}{20} + \frac{M^6}{20} \right]. \tag{5.6}$$

Using (P) property and (5.5), (5.6) we derive

Theorem 5.6 *Let $f : [m, M] \to \mathbb{R}$ be 3-times differentiable on $[m, M]$. Assume that f''' is bounded on $[m, M]$. Then*

$$\left\| f(A) - \left(\frac{1}{M-m} \int_m^M f(t)\,dt \right) 1_H - \left(\frac{f(M) - f(m)}{M-m} \right) \left(A - \left(\frac{m+M}{2} \right) 1_H \right) \right.$$
$$\left. - \left(\frac{f'(M) - f'(m)}{2(M-m)} \right) \left[A^2 - (m+M)A + \left(\frac{m^2 + M^2 + 4mM}{6} \right) 1_H \right] \right\|$$
$$\leq \frac{\|f'''\|_\infty}{(M-m)^3} Z(A), \tag{5.7}$$

where

$$Z(A) = \left[mMA^4 - \frac{1}{3}m^2M^3A + \frac{1}{3}m^3MA^2 - mM^2A^3 - \frac{1}{3}m^3M^2A + \right.$$
$$\frac{1}{3}mM^3A^2 + m^2M^2A^2 - m^2MA^3 - \frac{1}{2}mA^5 - \frac{1}{2}MA^5 + \frac{1}{6}A^6 + \frac{3}{4}m^2A^4 +$$

$$\frac{3}{4}M^2A^4+\left(\frac{1}{3}M^2m^4\right)1_H-\frac{2}{3}m^3A^3-\frac{2}{3}M^3A^3-\left(\frac{1}{3}M^3m^3\right)1_H+$$
$$\frac{5}{12}m^4A^2+\frac{5}{12}M^4A^2+\left(\frac{1}{3}M^4m^2\right)1_H-$$
$$\left(\frac{2}{15}Mm^5\right)1_H-\left(\frac{2}{15}mM^5\right)1_H-\frac{1}{6}m^5A-\frac{1}{6}M^5A+\left(\frac{m^6+M^6}{20}\right)1_H\Bigg]. \tag{5.8}$$

Let $f \in AC([m, M])$ (absolutely continuous functions on $[m, M]$), $0 < \alpha < 1$. Denote the right Caputo fractional derivative by $D_{t-}^{\alpha} f$ (see [4], p. 22) and the left Caputo fractional derivative by $D_{*t}^{\alpha} f$ (see [4], p. 78), $\forall\, t \in [m, M]$.

We need

Theorem 5.7 ([4], p. 44) *Let $0 < \alpha < 1$, $f \in AC([m, M])$, and $\left\|D_{t-}^{\alpha} f\right\|_{\infty,[m,t]}$, $\left\|D_{*t}^{\alpha} f\right\|_{\infty,[t,M]} < \infty$, $\forall\, t \in [m, M]$. Then*

$$\left|\frac{1}{M-m}\int_m^M f(z)\,dz - f(t)\right| \leq$$
$$\frac{1}{(M-m)\,\Gamma(\alpha+2)}\left\{\left\|D_{t-}^{\alpha} f\right\|_{\infty,[m,t]}(t-m)^{\alpha+1}+\left\|D_{*t}^{\alpha} f\right\|_{\infty,[t,M]}(M-t)^{\alpha+1}\right\} \leq \tag{5.9}$$

$$\frac{1}{\Gamma(\alpha+2)}\max\left\{\left\|D_{t-}^{\alpha} f\right\|_{\infty,[m,t]}, \left\|D_{*t}^{\alpha} f\right\|_{\infty,[t,M]}\right\}(M-m)^{\alpha}, \tag{5.10}$$

$\forall\, t \in [m, M]$.

By property (P) and Theorem 5.7 we derive

Theorem 5.8 *Let $0 < \alpha < 1$, $f \in AC([m, M])$, and there exists $K > 0$, such that*

$$\left\|D_{t-}^{\alpha} f\right\|_{\infty,[m,t]}, \left\|D_{*t}^{\alpha} f\right\|_{\infty,[t,M]} \leq K,\ \forall\, t \in [m, M]. \tag{5.11}$$

Then

$$\left|\left(\frac{1}{M-m}\int_m^M f(z)\,dz\right)1_H - f(A)\right| \leq$$
$$\frac{K}{(M-m)\,\Gamma(\alpha+2)}\left\{(A-m1_H)^{\alpha+1}+(M1_H-A)^{\alpha+1}\right\} \leq \tag{5.12}$$

$$\left(\frac{K}{\Gamma(\alpha+2)}(M-m)^{\alpha}\right)1_H. \tag{5.13}$$

We mention the Fink [10] inequality

Theorem 5.9 *Let $f^{(n-1)}$ be absolutely continuous on $[m, M]$ and $f^{(n)} \in L_{\infty}(m, M)$, $n \in \mathbb{N}$. Then*

$$\left| f(s) + \sum_{k=1}^{n-1} F_k(s) - \frac{n}{M-m} \int_m^M f(t)\, dt \right| \leq$$

$$\frac{\| f^{(n)} \|_{\infty}}{(n+1)!\,(M-m)} \left[(M-s)^{n+1} + (s-m)^{n+1} \right], \ \forall\, s \in [m, M], \tag{5.14}$$

where

$$F_k(s) := \left(\frac{n-k}{k!} \right) \left(\frac{f^{(k-1)}(m)(s-m)^k - f^{(k-1)}(M)(s-M)^k}{M-m} \right). \tag{5.15}$$

If $n = 1$, then $\sum_{k=1}^{n-1} = 0$.

Inequality (5.14) is sharp, in the sense that is attained by an optimal f for any $s \in [m, M]$.

By property (P) and Theorem 5.9 we obtain

Theorem 5.10 *Let $f^{(n-1)}$ be absolutely continuous on $[m, M]$ and $f^{(n)} \in L_{\infty}(m, M)$, $n \in \mathbb{N}$. Then*

$$\left| f(A) + \sum_{k=1}^{n-1} F_k(A) - \left(\frac{n}{M-m} \int_m^M f(t)\, dt \right) 1_H \right| \leq \tag{5.16}$$

$$\frac{\| f^{(n)} \|_{\infty}}{(n+1)!\,(M-m)} \left[(M 1_H - A)^{n+1} + (A - m 1_H)^{n+1} \right],$$

where

$$F_k(A) := \left(\frac{n-k}{k!} \right) \left(\frac{f^{(k-1)}(m)(A - m 1_H)^k - f^{(k-1)}(M)(A - M 1_H)^k}{M-m} \right). \tag{5.17}$$

If $n = 1$, then $\sum_{k=1}^{n-1} F_k(A) = 0_H$.

We use here the sequence $\{B_k(t), k \geq 0\}$ of Bernoulli polynomials which is uniquely determined by the following identities:

$$\begin{aligned} & B_k'(t) = kB_{k-1}(t), k \geq 1, B_0(t) = 1 \\ & \text{and} \\ & B_k(t+1) - B_k(t) = kt^{k-1}, k \geq 0. \end{aligned} \tag{5.18}$$

The values $B_k = B_k(0)$, $k \geq 0$ are the known Bernoulli numbers.

We mention

Theorem 5.11 ([3], p. 23, see also [6]) *Let $f : [m, M] \to \mathbb{R}$ be such that $f^{(n-1)}$, $n \in \mathbb{N}$, is a continuous function and $f^{(n)}(t)$ exists and is finite for all but a countable set of t in (m, M) and that $f^{(n)} \in L_\infty([m, M])$.*

Denote by

$$\Delta_n(s) := f(s) - \frac{1}{M-m}\int_m^M f(t)\,dt - \sum_{k=1}^{n-1}\frac{(M-m)^{k-1}}{k!}B_k\left(\frac{s-m}{M-m}\right)\left[f^{(k-1)}(M) - f^{(k-1)}(m)\right], \tag{5.19}$$

$\forall\, s \in [m, M]$.

Then

$$|\Delta_n(s)| \leq \frac{(M-m)^n}{n!}\left(\sqrt{\frac{(n!)^2}{(2n)!}|B_{2n}| + B_n^2\left(\frac{s-m}{M-m}\right)}\right)\left\|f^{(n)}\right\|_\infty, \tag{5.20}$$

$\forall\, n \in \mathbb{N}$; $\forall\, s \in [m, M]$.

Using the (P) property and Theorem 5.11 we derive:

Theorem 5.12 *All terms and assumptions as in Theorem 5.11. Denote by*

$$\Delta_n(A) := f(A) - \left(\frac{1}{M-m}\int_m^M f(t)\,dt\right)1_H - \sum_{k=1}^{n-1}\frac{(M-m)^{k-1}}{k!}B_k\left(\frac{A-m1_H}{M-m}\right)\left[f^{(k-1)}(M) - f^{(k-1)}(m)\right]. \tag{5.21}$$

Then

$$|\Delta_n(A)| \leq \frac{(M-m)^n}{n!}\left(\sqrt{\left(\frac{(n!)^2}{(2n)!}|B_{2n}|\right)1_H + B_n^2\left(\frac{A-m1_H}{M-m}\right)}\right)\left\|f^{(n)}\right\|_\infty, \tag{5.22}$$

$\forall\, n \in \mathbb{N}$.

Denote by (see [3], p. 24)

$$I_4(\lambda) := \begin{cases} \frac{16\lambda^5}{5} - 7\lambda^4 + \frac{14}{3}\lambda^3 - \lambda^2 + \frac{1}{30}, & 0 \leq \lambda \leq \frac{1}{2}, \\ -\frac{16\lambda^5}{5} + 9\lambda^4 - \frac{26\lambda^3}{3} + 3\lambda^2 - \frac{1}{10}, & \frac{1}{2} \leq \lambda \leq 1, \end{cases} \tag{5.23}$$

which is continuous in $\lambda \in [0, 1]$.

Also denote by

$$B := \left(\frac{A - m1_H}{M - m}\right)$$

and

$$I_4\left(\frac{A - m1_H}{M - m}\right) = I_4(B) =$$

$$\begin{cases} \frac{16}{5}B^5 - 7B^4 + \frac{14}{3}B^3 - B^2 + \frac{1}{30}1_H, & 0_H \leq B \leq \frac{1}{2}1_H, \\ -\frac{16}{5}B^5 + 9B^4 - \frac{26B^3}{3} + 3B^2 - \frac{1}{10}1_H, & \frac{1}{2}1_H \leq B \leq 1_H. \end{cases} \tag{5.24}$$

We mention

Theorem 5.13 ([3], p. 25) *All terms and assumptions as in Theorem 5.11, case of $n = 4$. For every $s \in [m, M]$ it holds*

$$|\Delta_4(s)| \leq \frac{(M - m)^4}{24} I_4(\lambda) \left\| f^{(4)} \right\|_\infty,$$

where $I_4(\lambda)$ is given by (5.23) with

$$\lambda = \frac{s - m}{M - m}. \tag{5.25}$$

Furthermore we have that

$$|\Delta_4(s)| \leq \frac{(M - m)^4}{720} \left\| f^{(4)} \right\|_\infty, \tag{5.26}$$

$\forall\, s \in [m, M]$.

Using property (P) and Theorem 5.13 we find

Theorem 5.14 *All terms and assumptions are according to Theorems 5.11–5.13. Then*

$$|\Delta_4(A)| \leq \frac{(M - m)^4}{24} I_4\left(\frac{A - m1_H}{M - m}\right) \left\| f^{(4)} \right\|_\infty, \tag{5.27}$$

where $I_4\left(\frac{A - m1_H}{M - m}\right)$ is given by (5.24).

Furthermore we have that

$$|\Delta_4(A)| \leq \left(\frac{(M-m)^4}{720}\left\|f^{(4)}\right\|_\infty\right) 1_H. \tag{5.28}$$

Next we follow [7].

Let $(P_n)_{n\in\mathbb{N}}$ be a harmonic sequence of polynomials, that is $P_n' = P_{n-1}$, $P_0 = 1$. Let $f : [m, M] \to \mathbb{R}$ be such that $f^{(n-1)}$ is absolutely continuous for some $n \in \mathbb{N}$. Setting

$$\overline{F_k} = \frac{(-1)^k (n-k)}{M-m}\left[P_k(m) f^{(k-1)}(m) - P_k(M) f^{(k-1)}(M)\right],\ k = 1, \ldots, n-1, \tag{5.29}$$

and

$$k(t, s) = \begin{cases} t - m, \text{ if } t \in [m, s] \\ t - M, \text{ if } t \in (s, M], \end{cases} \tag{5.30}$$

we get that

$$\frac{1}{n}\left[f(s) + \sum_{k=1}^{n-1}(-1)^k P_k(s) f^{(k)}(s) + \sum_{k=1}^{n-1}\overline{F_k}\right] - \frac{1}{M-m}\int_m^M f(t)\,dt = \tag{5.31}$$
$$\frac{(-1)^{n-1}}{n(M-m)}\int_m^M P_{n-1}(t)\,k(t,s)\,f^{(n)}(t)\,dt,$$

$\forall\, s \in [m, M]$. The above sums are defined to be zero for $n = 1$.

For the harmonic sequence of polynomials

$$P_k(t) = \frac{(t-s)^k}{k!},\ k \geq 0 \tag{5.32}$$

identity (5.31) collapses to the Fink identity, see [10].

We may rewrite generalized Fink identity (5.31) as follows:

$$f(s) = \sum_{k=1}^{n-1}(-1)^{k+1} P_k(s) f^{(k)}(s) + \tag{5.33}$$
$$\sum_{k=1}^{n-1}\frac{(-1)^k (n-k)}{M-m}\left[P_k(M) f^{(k-1)}(M) - P_k(m) f^{(k-1)}(m)\right] +$$
$$\frac{n}{M-m}\int_m^M f(t)\,dt + \frac{(-1)^{n+1}}{M-m}\int_m^M P_{n-1}(t)\,k(t,s)\,f^{(n)}(t)\,dt,$$

$\forall\, s \in [m, M]$, $n \in \mathbb{N}$, when $n = 1$ the above sums are zero.

Next we integrate the representation formula (5.33) against projections E_s to derive the operator representation formula:

$$f(A) = \sum_{k=1}^{n-1} (-1)^{k+1} P_k(A) f^{(k)}(A) + \tag{5.34}$$

$$\left[\sum_{k=1}^{n-1} \frac{(-1)^k (n-k)}{M-m} \left[P_k(M) f^{(k-1)}(M) - P_k(m) f^{(k-1)}(m) \right] + \right.$$

$$\left. \frac{n}{M-m} \int_m^M f(t)\,dt \right] 1_H + \frac{(-1)^{n+1}}{M-m} \int_{m-0}^M \left(\int_m^M P_{n-1}(t) k(t,s) f^{(n)}(t)\,dt \right) dE_s.$$

The sequence of polynomials

$$P_k(t) = \frac{1}{k!} \left(t - \frac{m+M}{2} \right)^k, \ k \geq 0, \tag{5.35}$$

is also harmonic.

We mention

Theorem 5.15 ([7]) *Let $f : [m, M] \to \mathbb{R}$ be such that $f^{(n-1)}$ is absolutely continuous for some $n \in \mathbb{N}$ and $f^{(n)} \in L_p([m, M])$, $1 \leq p \leq \infty$. Then*

$$\left| \left[f(s) + \sum_{k=1}^{n-1} (-1)^k P_k(s) f^{(k)}(s) + \sum_{k=1}^{n-1} \overline{F_k} \right] - \frac{n}{M-m} \int_m^M f(t)\,dt \right| \leq \tag{5.36}$$

$$\frac{1}{M-m} \| P_{n-1}(\cdot) k(\cdot, s) \|_{p',[m,M]} \left\| f^{(n)} \right\|_p,$$

where $\frac{1}{p} + \frac{1}{p'} = 1$.

We observe that

$$\int_m^M |P_{n-1}(t) k(t,s)|^{p'} dt \leq \| P_{n-1} \|_{\infty,[m,M]}^{p'} \int_m^M |k(t,s)|^{p'} dt = \tag{5.37}$$

$$\| P_{n-1} \|_{\infty,[m,M]}^{p'} \left[\int_m^s (t-m)^{p'} dt + \int_s^M (M-t)^{p'} dt \right] =$$

$$\| P_{n-1} \|_{\infty,[m,M]}^{p'} \left[\frac{(s-m)^{p'+1} + (M-s)^{p'+1}}{p'+1} \right].$$

Therefore we obtain

$$\| P_{n-1}(\cdot) k(\cdot, s) \|_{p',[m,M]} \leq \| P_{n-1} \|_{\infty,[m,M]} \left[\frac{(M-s)^{p'+1} + (s-m)^{p'+1}}{p'+1} \right]^{\frac{1}{p'}}. \tag{5.38}$$

Hence we have

Theorem 5.16 *Let $f : [m, M] \to \mathbb{R}$ be such that $f^{(n-1)}$ is absolutely continuous for some $n \in \mathbb{N}$ and $f^{(n)} \in L_p([m, M])$, $1 \leq p \leq \infty$. Then*

$$\left| \left(f(s) + \sum_{k=1}^{n-1} (-1)^k P_k(s) f^{(k)}(s) \right) + \left(\sum_{k=1}^{n-1} \overline{F_k} \right) - \left(\frac{n}{M-m} \int_m^M f(t)\,dt \right) \right| \leq$$
$$\left(\frac{\left\| f^{(n)} \right\|_p}{M-m} \left\| P_{n-1} \right\|_{\infty,[m,M]} \right) \left[\frac{(M-s)^{p'+1} + (s-m)^{p'+1}}{p'+1} \right]^{\frac{1}{p'}}, \tag{5.39}$$

$\forall\, s \in [m, M]$, where $\frac{1}{p} + \frac{1}{p'} = 1$.

We get the following operator inequality:

Theorem 5.17 *Let $f : [m, M] \to \mathbb{R}$ be such that $f^{(n-1)}$ is absolutely continuous for some $n \in \mathbb{N}$ and $f^{(n)} \in L_p([m, M])$, $1 \leq p \leq \infty$. Then*

$$\left| \left(f(A) + \sum_{k=1}^{n-1} (-1)^k P_k(A) f^{(k)}(A) \right) + \left(\sum_{k=1}^{n-1} \overline{F_k} \right) 1_H - \right.$$
$$\left. \left(\frac{n}{M-m} \int_m^M f(t)\,dt \right) 1_H \right| \leq$$
$$\left(\frac{\left\| f^{(n)} \right\|_p}{M-m} \left\| P_{n-1} \right\|_{\infty,[m,M]} \right) \left[\frac{(M1_H - A)^{p'+1} + (A - m1_H)^{p'+1}}{p'+1} \right]^{\frac{1}{p'}}, \tag{5.40}$$

where $\frac{1}{p} + \frac{1}{p'} = 1$.

Proof By (P) property and (5.39). ■

We give

Corollary 5.18 (to Theorem 5.16, see also [7]) *We have*

$$\left\| \left[f(s) + \sum_{k=1}^{n-1} \frac{(-1)^k}{k!} \left(s - \frac{m+M}{2} \right)^k f^{(k)}(s) + \right. \right.$$
$$\left. \sum_{k=1}^{n-1} \frac{(M-m)^{k-1}(n-k)}{k!2^k} \left[f^{(k-1)}(m) - (-1)^k f^{(k-1)}(M) \right] \right]$$

$$-\frac{n}{M-m}\int_m^M f(t)\,dt\Bigg| \leq$$

$$\left(\frac{\left\|f^{(n)}\right\|_p (M-m)^{n-2}}{2^{n-1}(n-1)!}\right)\left[\frac{(M-s)^{p'+1}+(s-m)^{p'+1}}{p'+1}\right]^{\frac{1}{p'}}, \tag{5.41}$$

$\forall\, s \in [m, M]$, where $\frac{1}{p}+\frac{1}{p'}=1$.

Proof Set $P_k(t)=\frac{1}{k!}\left(t-\frac{m+M}{2}\right)^k$, $k \geq 0$, in Theorem 5.16. ■

We finish with the operator inequality:

Corollary 5.19 (to Theorem 5.17) *We have*

$$\Bigg\|\Bigg[f(A)+\sum_{k=1}^{n-1}\frac{(-1)^k}{k!}\left(A-\left(\frac{m+M}{2}\right)1_H\right)^k f^{(k)}(A)+$$

$$\left(\sum_{k=1}^{n-1}\frac{(M-m)^{k-1}(n-k)}{k!2^k}\left[f^{(k-1)}(m)-(-1)^k f^{(k-1)}(M)\right]\right)1_H\Bigg]$$

$$-\left(\frac{n}{M-m}\int_m^M f(t)\,dt\right)1_H\Bigg\| \leq$$

$$\left(\frac{\left\|f^{(n)}\right\|_p (M-m)^{n-2}}{2^{n-1}(n-1)!}\right)\left[\frac{(M1_H-A)^{p'+1}+(A-m1_H)^{p'+1}}{p'+1}\right]^{\frac{1}{p'}}, \tag{5.42}$$

where $\frac{1}{p}+\frac{1}{p'}=1$.

Proof By Corollary 5.18 and (P) property. ■

References

1. G. Anastassiou, *Handbook of Analytic-Computational Methods in Applied Mathematics* (Chapman & Hall / CRC, Boca Raton, 2000)
2. G. Anastassiou, *Quantitative Approximations* (Chapman & Hall / CRC, Boca Raton, 2001)
3. G. Anastassiou, *Advanced Inequalities* (World Scientific, New Jersey, 2011)
4. G. Anastassiou, *Advances on Fractional Inequalities* (Springer, New York, 2011)
5. G. Anastassiou, Self adjoint operator Ostrowski type inequalities, J. Comput. Anal. Appl. (2016, accepted)
6. L.J. Dedic, M. Matic, J. Pečaric, On generalizations of Ostrowski inequality via some Euler-type identities. Math. Inequalities Appl. **3**(3), 337–353 (2000)
7. L.J. Dedic, J. Pečaric, N. Ujevic, On generalizations of Ostrowski inequality and some related results. Czechoslov. Math. J. **53**(128), 173–189 (2003)
8. S.S. Dragomir, *Inequalities for functions of selfadjoint operators on Hilbert Spaces* (2011), ajmaa.org/RGMIA/monographs/InFuncOp.pdf

9. S. Dragomir, *Operator Inequalities of Ostrowski and Trapezoidal Type* (Springer, New York, 2012)
10. A.M. Fink, Bounds on the deviation of a function from its averages. Czechoslov. Math. J. **42**(117), 289–310 (1992)
11. T. Furuta, J. Mićić Hot, J. Pečaric, Y. Seo, *Mond-Pečaric Method in Operator Inequalities. Inequalities for Bounded Selfadjoint Operators on a Hilobert Space* (Element, Zagreb, 2005)
12. G. Helmberg, *Introduction to Spectral Thery in Hilbert Space* (Wiley, New York, 1969)
13. A. Ostrowski, Über die Absolutabweichung einer differtentiebaren Funktion von ihrem Integralmittelwert. Comment. Math. Helv. **10**, 226–227 (1938)

Chapter 6
Integer and Fractional Self Adjoint Operator Opial Inequalities

We present here several integer and fractional self adjoint operator Opial type inequalities to many directions. These are based in the operator order over a Hilbert space. It follows [3].

6.1 Motivation

In 1960, Z. Opial ([10]) proved the following famous inequality that motivates our work here.

Let $f \in C^1([0,h])$ be such that $f(0) = f(h) = 0$, and $f(t) > 0$ in $(0,h)$. Then

$$\int_0^h |f(t) f'(t)| \, dt \leq \frac{h}{4} \int_0^h (f'(t))^2 \, dt.$$

The constant $\frac{h}{4}$ is the best.

In this chapter we present integer and fractional self adjoint operator Opial type inequalities on a Hilbert space in the operator order.

6.2 Background

Let A be a selfadjoint linear operator on a complex Hilbert space $(H; \langle \cdot, \cdot \rangle)$. The Gelfand map establishes a $*-$isometrically isomorphism Φ between the set $C(Sp(A))$ of all continuous functions defined on the spectrum of A, denoted $Sp(A)$, and the C^*-algebra $C^*(A)$ generated by A and the identity operator 1_H on H as follows (see e.g. [7, p. 3]):

For any $f, g \in C(Sp(A))$ and any $\alpha, \beta \in \mathbb{C}$ we have

(i) $\Phi(\alpha f + \beta g) = \alpha \Phi(f) + \beta \Phi(g)$;

G.A. Anastassiou, *Intelligent Comparisons II: Operator Inequalities and Approximations*, Studies in Computational Intelligence 699,
DOI 10.1007/978-3-319-51475-8_6

(ii) $\Phi(fg) = \Phi(f)\Phi(g)$ (the operation composition is on the right) and $\Phi(\overline{f}) = (\Phi(f))^*$;

(iii) $\|\Phi(f)\| = \|f\| := \sup_{t \in Sp(A)} |f(t)|$;

(iv) $\Phi(f_0) = 1_H$ and $\Phi(f_1) = A$, where $f_0(t) = 1$ and $f_1(t) = t$, for $t \in Sp(A)$.

With this notation we define

$$f(A) := \Phi(f), \text{ for all } f \in C(Sp(A)),$$

and we call it the continuous functional calculus for a selfadjoint operator A.

If A is a selfadjoint operator and f is a real valued continuous function on $Sp(A)$ then $f(t) \geq 0$ for any $t \in Sp(A)$ implies that $f(A) \geq 0$, i.e. $f(A)$ is a positive operator on H. Moreover, if both f and g are real valued continuous functions on $Sp(A)$ then the following important property holds:

(P) $f(t) \geq g(t)$ for any $t \in Sp(A)$, implies that $f(A) \geq g(A)$ in the operator order of $B(H)$. (the Banach algebra of all bounded linear operators from H into itself).

Equivalently, we use (see [6], pp. 7–8):

Let U be a selfadjoint operator on the complex Hilbert space $(H, \langle \cdot, \cdot \rangle)$ with the spectrum $Sp(U)$ included in the interval $[m, M]$ for some real numbers $m < M$ and $\{E_\lambda\}_\lambda$ be its spectral family.

Then for any continuous function $f : [m, M] \to \mathbb{C}$, it is well known that we have the following spectral representation in terms of the Riemann-Stieljes integral:

$$\langle f(U)x, y \rangle = \int_{m-0}^{M} f(\lambda)\, d(\langle E_\lambda x, y \rangle),$$

for any $x, y \in H$. The function $g_{x,y}(\lambda) := \langle E_\lambda x, y \rangle$ is of bounded variation on the interval $[m, M]$, and

$$g_{x,y}(m-0) = 0 \text{ and } g_{x,y}(M) = \langle x, y \rangle,$$

for any $x, y \in H$. Furthermore, it is known that $g_x(\lambda) := \langle E_\lambda x, x \rangle$ is increasing and right continuous on $[m, M]$.

We have also the formula

$$\langle f(U)x, x \rangle = \int_{m-0}^{M} f(\lambda)\, d(\langle E_\lambda x, x \rangle), \quad \forall\, x \in H.$$

As a symbol we can write

$$f(U) = \int_{m-0}^{M} f(\lambda)\, dE_\lambda.$$

Above, $m = \min \{\lambda | \lambda \in Sp(U)\} := \min Sp(U)$, $M = \max \{\lambda | \lambda \in Sp(U)\} := \max Sp(U)$. The projections $\{E_\lambda\}_{\lambda \in \mathbb{R}}$, are called the spectral family of A, with the properties:

(a) $E_\lambda \leq E_{\lambda'}$ for $\lambda \leq \lambda'$;
(b) $E_{m-0} = 0_H$ (zero operator), $E_M = 1_H$ (identity operator) and $E_{\lambda+0} = E_\lambda$ for all $\lambda \in \mathbb{R}$.

Furthermore

$$E_\lambda := \varphi_\lambda (U), \quad \forall \lambda \in \mathbb{R},$$

is a projection which reduces U, with

$$\varphi_\lambda (s) := \begin{cases} 1, & \text{for } -\infty < s \leq \lambda, \\ 0, & \text{for } \lambda < s < +\infty. \end{cases}$$

The spectral family $\{E_\lambda\}_{\lambda \in \mathbb{R}}$ determines uniquely the self-adjoint operator U and vice versa.

For more on the topic see [9], pp. 256–266, and for more details see there pp. 157–266. See also [5].

Some more basics are given (we follow [6], pp. 1–5):

Let $(H; \langle \cdot, \cdot \rangle)$ be a Hilbert space over $\mathbb{C}$. A bounded linear operator A defined on H is selfjoint, i.e., $A = A^*$, iff $\langle Ax, x \rangle \in \mathbb{R}$, $\forall x \in H$, and if A is selfadjoint, then

$$\|A\| = \sup_{x \in H : \|x\| = 1} |\langle Ax, x \rangle| .$$

Let A, B be selfadjoint operators on H. Then $A \leq B$ iff $\langle Ax, x \rangle \leq \langle Bx, x \rangle, \forall x \in H$.

In particular, A is called positive if $A \geq 0$.

Denote by

$$\mathcal{P} := \left\{ \varphi (s) := \sum_{k=0}^{n} \alpha_k s^k | n \geq 0, \alpha_k \in \mathbb{C}, 0 \leq k \leq n \right\}.$$

If $A \in \mathcal{B}(H)$ is selfadjoint, and $\varphi (s) \in \mathcal{P}$ has real coefficients, then $\varphi (A)$ is selfadjoint, and

$$\|\varphi (A)\| = \max\{|\varphi (\lambda)|, \lambda \in Sp(A)\}.$$

If φ is any function defined on $\mathbb{R}$ we define

$$\|\varphi\|_A := \sup\{|\varphi (\lambda)|, \lambda \in Sp(A)\}.$$

If A is selfadjoint operator on Hilbert space H and φ is continuous and given that $\varphi (A)$ is selfadjoint, then $\|\varphi (A)\| = \|\varphi\|_A$. And if φ is a continuous real valued function so it is $|\varphi|$, then $\varphi (A)$ and $|\varphi| (A) = |\varphi (A)|$ are selfadjoint operators (by [6], p. 4, Theorem 7).

Hence it holds

$$\| |\varphi(A)| \| = \| |\varphi| \|_A = \sup \{ \| \varphi(\lambda) \| , \lambda \in Sp(A) \}$$
$$= \sup \{ |\varphi(\lambda)| , \lambda \in Sp(A) \} = \|\varphi\|_A = \|\varphi(A)\|,$$

that is

$$\| |\varphi(A)| \| = \|\varphi(A)\|.$$

For a selfadjoint operator $A \in \mathcal{B}(H)$ which is positive, there exists a unique positive selfadjoint operator $B := \sqrt{A} \in \mathcal{B}(H)$ such that $B^2 = A$, that is $\left(\sqrt{A}\right)^2 = A$. We call B the square root of A.

Let $A \in \mathcal{B}(H)$, then A^*A is selfadjoint and positive. Define the "operator absolute value" $|A| := \sqrt{A^*A}$. If $A = A^*$, then $|A| = \sqrt{A^2}$.

For a continuous real valued function φ we observe the following:

$$|\varphi(A)| \text{ (the functional absolute value) } = \int_{m-0}^{M} |\varphi(\lambda)| \, dE_\lambda =$$

$$\int_{m-0}^{M} \sqrt{(\varphi(\lambda))^2} dE_\lambda = \sqrt{(\varphi(A))^2} = |\varphi(A)| \text{ (operator absolute value)},$$

where A is a selfadjoint operator.

That is we have

$$|\varphi(A)| \text{ (functional absolute value) } = |\varphi(A)| \text{ (operator absolute value)}.$$

6.3 Main Results

Let A be a selfadjoint operator in the Hilbert space H with the spectrum $Sp(A) \subseteq [m, M]$, $m < M$; $m, M \in \mathbb{R}$.

In the next we obtain Opial type inequalities, both integer and fractional cases, in the operator order of $\mathcal{B}(H)$ (the Banach algebra of all bounded linear operators from H into itself).

Let the real valued function $f \in C([m, M])$, and we consider

$$g(t) = \int_m^t f(z) \, dz, \quad \forall t \in [m, M], \tag{6.1}$$

then $g \in C([m, M])$.

We denote by

$$\int_{m1_H}^{A} f := \Phi(g) = g(A). \tag{6.2}$$

We understand and write that $(r > 0)$

$$g^r(A) = \Phi(g^r) =: \left(\int_{m1_H}^{A} f\right)^r.$$

Clearly $\left(\int_{m1_H}^{A} f\right)^r$ is a self adjoint operator on H, for any $r > 0$.

All of our functions in this chapter will be real valued. From [4] we mention the following basic version of Opial inequality:

Theorem 6.1 *Let $f \in C^1([m, M])$ with $f(m) = 0$. Then*

$$\int_m^{\lambda} |f(t)| \left|f'(t)\right| dt \le \left(\frac{\lambda - m}{2}\right) \int_m^{\lambda} \left(f'(t)\right)^2 dt, \ \forall \lambda \in [m, M]. \tag{6.3}$$

When $f(t) = t - m$, $t \in [m, M]$, inequality (6.3) becomes equality.

By applying properties (P) and (ii) to (6.3) we obtain

Theorem 6.2 *Let $f \in C^1([m, M])$ with $f(m) = 0$. Then*

$$\int_{m1_H}^{A} \left|ff'\right| \le \frac{1}{2}(A - m1_H) \left(\int_{m1_H}^{A} \left(f'\right)^2\right). \tag{6.4}$$

We mention

Theorem 6.3 ([4]) *Let $f \in C^1([m, M])$ with $f(m) = 0$, and $1 \le p \le 2$. Then*

$$\int_m^{\lambda} |f(t)|^p \left|f'(t)\right|^p dt \le K(p)(\lambda - m) \left(\int_m^{\lambda} \left(f'(t)\right)^2 dt\right)^p, \ \forall \lambda \in [m, M], \tag{6.5}$$

where

$$K(p) = \begin{cases} \frac{1}{2}, \ p = 1, \\ \frac{4}{\pi^2}, \ p = 2, \\ \frac{2-p}{2p}\left(\frac{1}{p}\right)^{2p-2} I^{-p}, \ 1 < p < 2, \end{cases} \tag{6.6}$$

with

$$I = \int_0^1 \left\{1 + \frac{2(p-1)}{2-p} z\right\}^{-2} \{1 + (p-1)z\}^{\frac{1}{p}-1} dz.$$

For $p = 1$, equality holds in (6.5) only for f linear.

By applying properties (P) and (ii) to (6.5) we derive

Theorem 6.4 *Here all are as in Theorem 6.3. It holds*

$$\int_{m1_H}^{A} \left|ff'\right|^p \le K(p)(A - m1_H) \left(\int_{m1_H}^{A} \left(f'\right)^2\right)^p. \tag{6.7}$$

We mention

Theorem 6.5 ([8]) *Let $f \in C^1([m, M])$ with $f(m) = 0$, and $p, q \geq 1$. Then*

$$\int_m^\lambda |f(t)|^p |f'(t)|^q dt \leq \left(\frac{q}{p+q}\right)(\lambda - m)^p \int_m^\lambda |f'(t)|^{p+q} dt, \ \forall \lambda \in [m, M]. \tag{6.8}$$

By applying properties (P) and (ii) to (6.8) we find

Theorem 6.6 *Let $f \in C^1([m, M])$ with $f(m) = 0$, and $p, q \geq 1$. Then*

$$\int_{m1_H}^A |f|^p |f'|^q \leq \left(\frac{q}{p+q}\right)(A - m1_H)^p \left(\int_{m1_H}^A |f'|^{p+q}\right). \tag{6.9}$$

We mention

Theorem 6.7 ([12]) *Let $p > -1$. Let $f \in C^1([m, M])$, and $f(m) = 0$. Then*

$$\int_m^\lambda t^p |f(t) f'(t)| dt \leq \frac{1}{2\sqrt{p+1}} \int_m^\lambda \left(\lambda^{p+1} - mt^p\right)\left(f'(t)\right)^2 dt \tag{6.10}$$

$$\leq \frac{1}{2\sqrt{p+1}} \int_m^\lambda \left(M^{p+1} - mt^p\right)\left(f'(t)\right)^2 dt, \ \forall \lambda \in [m, M]. \tag{6.11}$$

(inequality (6.11) is our derivation).

By applying properties (P) and (ii) to (6.10), (6.11) we obtain

Theorem 6.8 *Let $p > -1$. Let $f \in C^1([m, M])$ and $f(m) = 0$. Then*

$$\int_{m1_H}^A (id)^p |ff'| \leq \frac{1}{2\sqrt{p+1}} \left(\int_{m1_H}^A \left(M^{p+1} - m(id)^p\right)\left(f'\right)^2\right). \tag{6.12}$$

We mention

Theorem 6.9 ([1], p. 20) *Let $q(t)$ be positive continuous and non-increasing function on $[m, M]$. Further, let $f \in C^1([m, M])$, and $f(m) = 0$. Let $l \geq 0$, $w \geq 1$. Then*

$$\int_m^\lambda q(t) |f(t)|^l |f'(t)|^w dt \leq \left(\frac{w}{l+w}\right)(\lambda - m)^l \int_m^\lambda q(t) |f'(t)|^{l+w} dt, \tag{6.13}$$

$\forall \lambda \in [m, M]$.

By applying property (P) and (ii) to (6.13) we obtain

Theorem 6.10 *All as in Theorem 6.9. Then*

$$\int_{m1_H}^{A} q\,|f|^l\,|f'|^w \leq \left(\frac{w}{l+w}\right)(A - m1_H)^l \int_{m1_H}^{A} q\,|f'|^{l+w}. \tag{6.14}$$

We mention

Theorem 6.11 (see [1], p. 68) *Let $q(t)$ positive, continuous and non-increasing on $[m, M]$. Further let $f_1, f_2 \in C^1([m, M])$ with $f_1(m) = f_2(m) = 0$. Let $l \geq 0$, $w \geq 1$. Then*

$$\int_m^{\lambda} q(t)\,|f_1(t)\,f_2(t)|^l \left[|f_1(t)\,f_2'(t)|^w + |f_1'(t)\,f_2(t)|^w\right] dt \leq$$

$$\frac{w}{2(l+w)}(\lambda - m)^{2l+w} \int_m^{\lambda} q(t)\left[\left(f_1'(t)\right)^{2(l+w)} + \left(f_2'(t)\right)^{2(l+w)}\right] dt, \tag{6.15}$$

$\forall\,\lambda \in [m, M]$.

By applying property (P) and (ii) to (6.15) we obtain

Theorem 6.12 *All as in Theorem 6.11. Then*

$$\int_{m1_H}^{A} q\,|f_1 f_2|^l \left[|f_1 f_2'|^w + |f_1' f_2|^w\right] \leq$$

$$\frac{w}{2(l+w)}(A - m1_H)^{2l+w} \int_{m1_H}^{A} q\left[\left(f_1'\right)^{2(l+w)} + \left(f_2'\right)^{2(l+w)}\right]. \tag{6.16}$$

We mention

Theorem 6.13 ([11], p. 308) *Let $f \in C^n([m, M])$, $n \in \mathbb{N}$, $f^{(i)}(m) = 0$, for $i = 0, 1, 2, ..., n-1$. Then*

$$\int_m^{\lambda} \left|f(t)\,f^{(n)}(t)\right| dt \leq \frac{(\lambda - m)^n}{2} \int_m^{\lambda} \left(f^{(n)}(t)\right)^2 dt, \ \forall\,\lambda \in [m, M]. \tag{6.17}$$

Using properties (P) and (ii) on (6.17) we derive

Theorem 6.14 *All as in Theorem 6.13. Then*

$$\int_{m1_H}^{A} \left|f \cdot f^{(n)}\right| \leq \frac{(A - m1_H)^n}{2}\left(\int_{m1_H}^{A} \left(f^{(n)}\right)^2\right). \tag{6.18}$$

We mention from [11], p. 309

Theorem 6.15 *Let $f_1, f_2 \in C^n([m, M])$ such that $f_1^{(k)}(m) = f_2^{(k)}(m) = 0$, for $k = 0, 1, ..., n-1$, $n \in \mathbb{N}$. Then*

$$\int_m^{\lambda}\left[\left|f_1(t)\, f_2^{(n)}(t)\right| + \left|f_2(t)\, f_1^{(n)}(t)\right|\right] dt \le$$

$$B(\lambda - m)^n \int_m^{\lambda}\left[\left(f_1^{(n)}(t)\right)^2 + \left(f_2^{(n)}(t)\right)^2\right] dt, \ \forall\, \lambda \in [m, M], \tag{6.19}$$

where

$$B = \frac{1}{2n!}\left(\frac{n}{2n-1}\right)^{\frac{1}{2}}. \tag{6.20}$$

Using (6.19) and properties (P) and (ii) we obtain

Theorem 6.16 *All as in Theorem 6.15. Then*

$$\int_{m1_H}^{A}\left[\left|f_1 f_2^{(n)}\right| + \left|f_2 f_1^{(n)}\right|\right] \le$$

$$B(A - m1_H)^n \left(\int_{m1_H}^{A}\left(\left(f_1^{(n)}\right)^2 + \left(f_2^{(n)}\right)^2\right)\right). \tag{6.21}$$

Here we follow [2], p. 8.

Definition 6.17 *Let $\nu > 0$, $n := [\nu]$ (integral part), and $\alpha := \nu - n$ ($0 < \alpha < 1$). Let $f \in C([m, M])$ and define*

$$\left(J_{\nu}^{m} f\right)(z) = \frac{1}{\Gamma(\nu)} \int_m^z (z - t)^{\nu - 1} f(t)\, dt, \tag{6.22}$$

all $m \le z \le M$, where Γ is the gamma function, the generalized Riemann-Liouville integral. We define the subspace $C_m^{\nu}([m, M])$ of $C^n([m, M])$:

$$C_m^{\nu}([m, M]) := \left\{f \in C^n([m, M]) : J_{1-\alpha}^{m} f^{(n)} \in C^1([m, M])\right\}. \tag{6.23}$$

So let $f \in C_m^{\nu}([m, M])$; we define the generalized ν-fractional derivative (of Canavati type) of f over $[m, M]$ as

$$D_m^{\nu} f := \left(J_{1-\alpha}^{m} f^{(n)}\right)'. \tag{6.24}$$

Notice that

$$\left(J_{1-\alpha}^{m} f^{(n)}\right)(z) = \frac{1}{\Gamma(1-\alpha)} \int_m^z (z - t)^{-\alpha} f^{(n)}(t)\, dt \tag{6.25}$$

exists for $f \in C_m^{\nu}([m, M])$, all $m \le z \le M$.

Also notice that $D_m^{\nu} f \in C([m, M])$.

We need

Theorem 6.18 ([2], p. 15) *Let* $f \in C_m^\nu([m, M])$, $\nu \geq 1$ *and* $f^{(i)}(m) = 0$, $i = 0, 1, \ldots, n-1$, $n := [\nu]$. *Here* $\lambda \in [m, M]$, *and* $l = 1, \ldots, n-1$. *Let* $p, q > 1 : \frac{1}{p} + \frac{1}{q} = 1$. *Then*

$$\int_m^\lambda \left|f^{(l)}(w)\right| \left|\left(D_m^\nu f\right)(w)\right| dw \leq$$

$$\frac{2^{-\frac{1}{q}} (\lambda - m)^{\frac{(\nu p - lp - p + 2)}{p}}}{\Gamma(\nu - l)\left((\nu p - lp - p + 1)(\nu p - lp - p + 2)\right)^{\frac{1}{p}}} \left(\int_m^\lambda \left|\left(D_m^\nu f\right)(w)\right|^q dw\right)^{\frac{2}{q}}. \tag{6.26}$$

Using (6.26), properties (P) and (ii) we get

Theorem 6.19 *All as in Theorem 6.18. Then*

$$\int_{m1_H}^A \left|f^{(l)}\right| \left|\left(D_m^\nu f\right)\right| \leq$$

$$\frac{2^{-\frac{1}{q}} (A - m1_H)^{\frac{(\nu p - lp - p + 2)}{p}}}{\Gamma(\nu - l)\left((\nu p - lp - p + 1)(\nu p - lp - p + 2)\right)^{\frac{1}{p}}} \left(\int_{m1_H}^A \left|\left(D_m^\nu f\right)\right|^q\right)^{\frac{2}{q}}. \tag{6.27}$$

We need

Theorem 6.20 ([2], p. 26) *Let* $\gamma_1, \gamma_2 \geq 0$, $\nu \geq 1$ *be such that* $\nu - \gamma_1, \nu - \gamma_2 \geq 1$ *and* $f \in C_m^\nu([m, M])$ *with* $f^{(i)}(m) = 0$, $i = 0, 1, \ldots, n-1$, $n := [\nu]$. *Here* $\lambda \in [m, M]$. *Let* q *be a nonnegative continuous functions on* $[m, M]$. *Denote*

$$Q(\lambda) := \left(\int_m^\lambda (q(w))^2 dw\right)^{\frac{1}{2}}, \ \forall \lambda \in [m, M]. \tag{6.28}$$

Then

$$\int_m^\lambda q(w) \left|D_m^{\gamma_1}(f)(w)\right| \left|D_m^{\gamma_2}(f)(w)\right| dw \leq$$

$$K(q, \gamma_1, \gamma_2, \nu, \lambda, m) \left(\int_m^\lambda \left(D_m^\nu f(w)\right)^2 dw\right), \tag{6.29}$$

where

$$K(q, \gamma_1, \gamma_2, \nu, \lambda, m) := \frac{Q(\lambda)}{\sqrt[3]{6}} \frac{1}{\Gamma(\nu - \gamma_1)\Gamma(\nu - \gamma_2)}.$$

$$\frac{(\lambda - m)^{2\nu - \gamma_1 - \gamma_2 - \frac{1}{2}}}{\left(\nu - \gamma_1 - \frac{5}{6}\right)^{\frac{1}{6}} \left(\nu - \gamma_2 - \frac{5}{6}\right)^{\frac{1}{6}} \left(4\nu - 2\gamma_1 - 2\gamma_2 - \frac{7}{3}\right)^{\frac{1}{2}}}. \tag{6.30}$$

Using (6.30) and Remark 3.4 of [2], p. 26, and properties (P) and (ii) to obtain

Theorem 6.21 *All terms and assumptions as in Theorem 6.20. Then*

$$\int_{m1_H}^{A} q \left| D_m^{\gamma_1}(f) \right| \left| D_m^{\gamma_2}(f) \right| \leq$$

$$K(q, \gamma_1, \gamma_2, \nu, A, m) \left(\int_{m1_H}^{A} \left(D_m^{\nu} f \right)^2 \right), \tag{6.31}$$

where

$$K(q, \gamma_1, \gamma_2, \nu, A, m) := \frac{Q(A)}{\sqrt[3]{6}} \frac{1}{\Gamma(\nu - \gamma_1) \Gamma(\nu - \gamma_2)}.$$

$$\frac{(A - m1_H)^{2\nu - \gamma_1 - \gamma_2 - \frac{1}{2}}}{\left(\nu - \gamma_1 - \frac{5}{6}\right)^{\frac{1}{6}} \left(\nu - \gamma_2 - \frac{5}{6}\right)^{\frac{1}{6}} \left(4\nu - 2\gamma_1 - 2\gamma_2 - \frac{7}{3}\right)^{\frac{1}{2}}}. \tag{6.32}$$

We need

Theorem 6.22 ([2], p. 30) *Let $\gamma \geq 0$, $\nu \geq 1$, $\nu - \gamma \geq 1$, let q be a nonnegative continuous function on $[m, M]$. Let $f \in C_m^{\nu}([m, M])$ with $f^{(i)}(m) = 0$, $i = 0, 1, ..., n-1$, $n := [\nu]$. Let $\lambda \in [m, M]$. Call*

$$Q(\lambda) := \left(\int_m^{\lambda} (q(w))^2 (w - m)^{2\nu - 2\gamma - 1} dw \right)^{\frac{1}{2}}, \tag{6.33}$$

and

$$K(q, \gamma, \nu, \lambda, m) := \frac{Q(\lambda)}{\sqrt{2(2\nu - 2\gamma - 1)}\Gamma(\nu - \gamma)}. \tag{6.34}$$

Then

$$\int_m^{\lambda} q(w) \left| D_m^{\gamma} f(w) \right| \left| D_m^{\nu} f(w) \right| dw \leq K(q, \gamma, \nu, \lambda, m) \left(\int_m^{\lambda} \left(\left(D_m^{\nu} f \right)(w) \right)^2 dw \right). \tag{6.35}$$

Using (6.33)–(6.35) and properties (P) and (ii) we derive

Theorem 6.23 *All as in Theorem 6.22. Denote by*

$$K(q, \gamma, \nu, A, m) := \frac{Q(A)}{\sqrt{2(2\nu - 2\gamma - 1)}\Gamma(\nu - \gamma)}. \tag{6.36}$$

Then

$$\int_{m1_H}^{A} q \left| D_m^{\gamma} f \right| \left| D_m^{\nu} f \right| \leq K(q, \gamma, \nu, A, m) \left(\int_{m1_H}^{A} \left(\left(D_m^{\nu} f \right) \right)^2 \right). \tag{6.37}$$

We need

Theorem 6.24 ([2], p. 92) *Let* $\nu \geq 1$, $\gamma_1, \gamma_2 \geq 0$, *such that* $\nu - \gamma_1 \geq 1$, $\nu - \gamma_2 \geq 1$, *and* $f_1, f_2 \in C_m^{\nu}([m, M])$ *with* $f_1^{(i)}(m) = f_2^{(i)}(m) = 0$, $i = 0, 1, ..., n-1$, $n := [\nu]$. *Here* $\lambda \in [m, M]$. *Let* $\lambda_\alpha, \lambda_\beta, \lambda_\nu \geq 0$. *Set*

$$\rho(\lambda) := \frac{(\lambda - m)^{(\nu\lambda_\alpha - \gamma_1\lambda_\alpha + \nu\lambda_\beta - \gamma_2\lambda_\beta + 1)}}{(\nu\lambda_\alpha - \gamma_1\lambda_\alpha + \nu\lambda_\beta - \gamma_2\lambda_\beta + 1)(\Gamma(\nu - \gamma_1 + 1))^{\lambda_\alpha}(\Gamma(\nu - \gamma_2 + 1))^{\lambda_\beta}}. \tag{6.38}$$

Then

$$\int_m^\lambda \left[\left| \left(D_m^{\gamma_1} f_1\right)(w) \right|^{\lambda_\alpha} \left| \left(D_m^{\gamma_2} f_2\right)(w) \right|^{\lambda_\beta} \left| \left(D_m^{\nu} f_1\right)(w) \right|^{\lambda_\nu} + \right.$$

$$\left. \left| \left(D_m^{\gamma_2} f_1\right)(w) \right|^{\lambda_\beta} \left| \left(D_m^{\gamma_1} f_2\right)(w) \right|^{\lambda_\alpha} \left| \left(D_m^{\nu} f_2\right)(w) \right|^{\lambda_\nu} \right] dw \leq$$

$$\frac{\rho(\lambda)}{2} \left[\left\| D_m^{\nu} f_1 \right\|_\infty^{2(\lambda_\alpha + \lambda_\nu)} + \left\| D_m^{\nu} f_1 \right\|_\infty^{2\lambda_\beta} + \left\| D_m^{\nu} f_2 \right\|_\infty^{2\lambda_\beta} + \left\| D_m^{\nu} f_2 \right\|_\infty^{2(\lambda_\alpha + \lambda_\nu)} \right], \tag{6.39}$$

all $m \leq \lambda \leq M$.

Using (6.39) and properties (P) and (ii) we derive

Theorem 6.25 *All here as in Theorem 6.24. Set*

$$\rho(A) := \frac{(A - m1_H)^{(\nu\lambda_\alpha - \gamma_1\lambda_\alpha + \nu\lambda_\beta - \gamma_2\lambda_\beta + 1)}}{(\nu\lambda_\alpha - \gamma_1\lambda_\alpha + \nu\lambda_\beta - \gamma_2\lambda_\beta + 1)(\Gamma(\nu - \gamma_1 + 1))^{\lambda_\alpha}(\Gamma(\nu - \gamma_2 + 1))^{\lambda_\beta}}. \tag{6.40}$$

Then

$$\int_{m1_H}^A \left[\left| \left(D_m^{\gamma_1} f_1\right) \right|^{\lambda_\alpha} \left| \left(D_m^{\gamma_2} f_2\right) \right|^{\lambda_\beta} \left| \left(D_m^{\nu} f_1\right) \right|^{\lambda_\nu} + \right.$$

$$\left. \left| \left(D_m^{\gamma_2} f_1\right) \right|^{\lambda_\beta} \left| \left(D_m^{\gamma_1} f_2\right) \right|^{\lambda_\alpha} \left| \left(D_m^{\nu} f_2\right) \right|^{\lambda_\nu} \right] \leq$$

$$\frac{\rho(A)}{2} \left[\left\| D_m^{\nu} f_1 \right\|_\infty^{2(\lambda_\alpha + \lambda_\nu)} + \left\| D_m^{\nu} f_1 \right\|_\infty^{2\lambda_\beta} + \left\| D_m^{\nu} f_2 \right\|_\infty^{2\lambda_\beta} + \left\| D_m^{\nu} f_2 \right\|_\infty^{2(\lambda_\alpha + \lambda_\nu)} \right]. \tag{6.41}$$

We give

Definition 6.26 ([2], p. 270) *Let* $\nu > 0$, $n := \lceil \nu \rceil$ *(ceiling of* ν*),* $f \in AC^n([m, M])$ *(i.e.* $f^{(n-1)}$ *is absolutely continuous on* $[m, M]$*, that is in* $AC([m, M])$*). We define the Caputo fractional derivative*

$$\left(D_{*m}^{\nu} f\right)(z) := \frac{1}{\Gamma(n - \nu)} \int_m^z (z - t)^{n - \nu - 1} f^{(n)}(t)\, dt, \tag{6.42}$$

which exists almost everywhere for $z \in [m, M]$.

Notice that $D^0_{*m} f = f$, *and* $D^n_{*m} f = f^{(n)}$.

We mention

Theorem 6.27 ([2], p. 397) *Let* $\nu \geq \gamma + 1$, $\gamma \geq 0$. *Call* $n := \lceil \nu \rceil$ *and assume* $f \in C^n([m, M])$ *such that* $f^{(k)}(m) = 0$, $k = 0, 1, ..., n-1$. *Let* $p, q > 1 : \frac{1}{p} + \frac{1}{q} = 1$, $m \leq \lambda \leq M$. *Then*

$$\int_m^\lambda \left| \left(D^\gamma_{*m} f \right)(w) \right| \left| \left(D^\nu_{*m} f \right)(w) \right| dw \leq$$

$$\frac{(\lambda - m)^{\frac{(p\nu - p\gamma - p + 2)}{p}}}{\left(\sqrt[q]{2}\right) \Gamma(\nu - \gamma) \left((p\nu - p\gamma - p + 1)(p\nu - p\gamma - p + 2)\right)^{\frac{1}{p}}} \left(\int_m^\lambda \left| D^\nu_{*m} f(w) \right|^q dw \right)^{\frac{2}{q}}. \quad (6.43)$$

Note: By Proposition 15.114 ([2], p. 388) we have that $D^\nu_{*m} f$, $D^\gamma_{*m} f \in C([m, M])$.

Using (6.43) and Properties (P) and (ii) we give

Theorem 6.28 *All as in Theorem 6.27. Then*

$$\int_{m1_H}^A \left| \left(D^\gamma_{*m} f \right) \right| \left| \left(D^\nu_{*m} f \right) \right| \leq$$

$$\frac{(A - m1_H)^{\frac{(p\nu - p\gamma - p + 2)}{p}}}{\left(\sqrt[q]{2}\right) \Gamma(\nu - \gamma) \left((p\nu - p\gamma - p + 1)(p\nu - p\gamma - p + 2)\right)^{\frac{1}{p}}} \left(\int_{m1_H}^A \left| D^\nu_{*m} f \right|^q \right)^{\frac{2}{q}}. \quad (6.44)$$

We need

Theorem 6.29 ([2], p. 398) *Let* $\nu \geq 2$, $k \geq 0$, $\nu \geq k + 2$. *Call* $n := \lceil \nu \rceil$ *and* $f \in C^n([m, M]) : f^{(j)}(m) = 0$, $j = 0, 1, ..., n-1$. *Let* $p, q > 1 : \frac{1}{p} + \frac{1}{q} = 1$, $m \leq \lambda \leq M$. *Then*

$$\int_m^\lambda \left| \left(D^k_{*m} f \right)(w) \right| \left| \left(D^{k+1}_{*m} f \right)(w) \right| dw \leq$$

$$\frac{(\lambda - m)^{\frac{2(p\nu - pk - p + 1)}{p}}}{2 \left(\Gamma(\nu - k) \right)^2 (p\nu - pk - p + 1)^{\frac{2}{p}}} \left(\int_m^\lambda \left| D^\nu_{*m} f(w) \right|^q dw \right)^{\frac{2}{q}}. \quad (6.45)$$

Using (6.45) and Properties (P) and (ii) we find

Theorem 6.30 *All as in Theorem 6.29. Then*

$$\int_{m1_H}^A \left| \left(D^k_{*m} f \right) \right| \left| \left(D^{k+1}_{*m} f \right) \right| \leq$$

$$\frac{(A - m1_H)^{\frac{2(p\nu - pk - p + 1)}{p}}}{2 \left(\Gamma(\nu - k) \right)^2 (p\nu - pk - p + 1)^{\frac{2}{p}}} \left(\int_{m1_H}^A \left| D^\nu_{*m} f \right|^q \right)^{\frac{2}{q}}. \quad (6.46)$$

We need

Theorem 6.31 ([2], p. 399) *Let* $\gamma_i \geq 0$, $\nu \geq 1$, $\nu - \gamma_i \geq 1$; $i = 1, ..., l$, $n := \lceil \nu \rceil$, *and* $f \in C^n([m, M])$ *such that* $f^{(k)}(m) = 0$, $k = 0, 1, ..., n-1$. *Here* $m \leq \lambda \leq M$; $q_1(\lambda), q_2(\lambda)$ *continuous functions on* $[m, M]$ *such that* $q_1(\lambda) \geq 0$, $q_2(\lambda) > 0$ *on* $[m, M]$, *and* $r_i > 0 : \sum_{i=1}^{l} r_i = r$. *Let* $s_1, s_1' > 1 : \frac{1}{s_1} + \frac{1}{s_1'} = 1$ *and* $s_2, s_2' > 1 : \frac{1}{s_2} + \frac{1}{s_2'} = 1$, *and* $p > s_2$.

Denote by

$$Q_1(\lambda) := \left(\int_m^{\lambda} (q_1(w))^{s_1'} dw \right)^{\frac{1}{s_1'}} \tag{6.47}$$

and

$$Q_2(\lambda) := \left(\int_m^{\lambda} (q_2(w))^{\frac{-s_2'}{p}} dw \right)^{\frac{r}{s_2'}}, \tag{6.48}$$

$$\sigma := \frac{p - s_2}{p s_2}. \tag{6.49}$$

Then

$$\int_m^{\lambda} q_1(w) \prod_{i=1}^{l} \left| D_{*m}^{\gamma_i} f(w) \right|^{r_i} dw \leq$$

$$Q_1(\lambda) Q_2(\lambda) \prod_{i=1}^{l} \left\{ \frac{\sigma^{r_i \sigma}}{(\Gamma(\nu - \gamma_i))^{r_i} (\nu - \gamma_i - 1 + \sigma)^{r_i \sigma}} \right\}.$$

$$\frac{(\lambda - m)^{\left(\sum_{i=1}^{l}(\nu - \gamma_i - 1) r_i + \sigma r\right) + \frac{1}{s_1}}}{\left(\left(\sum_{i=1}^{l} (\nu - \gamma_i - 1) r_i s_1 \right) + r s_1 \sigma + 1 \right)^{\frac{1}{s_1}}} \left(\int_m^{\lambda} q_2(w) \left| D_{*m}^{\nu} f(w) \right|^p dw \right)^{\frac{r}{p}}. \tag{6.50}$$

Using (6.50) and properties (P) and (ii) we obtain

Theorem 6.32 *All here as in Theorem 6.31. Set*

$$Q_1(A) := \left(\int_{m1_H}^{A} (q_1)^{s_1'} \right)^{\frac{1}{s_1'}} \tag{6.51}$$

and

$$Q_2(A) := \left(\int_{m1_H}^{A} (q_2)^{\frac{-s_2'}{p}} \right)^{\frac{r}{s_2'}}. \tag{6.52}$$

Then

$$\int_{m1_H}^{A} q_1 \prod_{i=1}^{l} \left| D_{*m}^{\gamma_i} f \right|^{r_i} \leq$$

$$Q_1(A)\, Q_2(A) \prod_{i=1}^{l} \left\{ \frac{\sigma^{r_i \sigma}}{(\Gamma(\nu - \gamma_i))^{r_i} (\nu - \gamma_i - 1 + \sigma)^{r_i \sigma}} \right\} \cdot$$

$$\frac{(A - m1_H)^{\left(\sum_{i=1}^{l} (\nu - \gamma_i - 1) r_i + \sigma r\right) + \frac{1}{s_1}}}{\left(\left(\sum_{i=1}^{l} (\nu - \gamma_i - 1) r_i s_1\right) + r s_1 \sigma + 1\right)^{\frac{1}{s_1}}} \left(\int_{m1_H}^{A} q_2 \left| D_{*m}^{\nu} f \right|^{p} \right)^{\frac{r}{p}}. \tag{6.53}$$

One can give many more operator Opial type (both integer and fractional) inequalities.

We choose to stop here.

References

1. R.P. Agarwal, P.Y.H. Pang, *Opial Inequalities with Applications in Diferential and Difference Equations* (Kluwer Acadmic Publisher, Dordrecht, Boston, London, 1995)
2. G. Anastassiou, *Fractional Differentiation Inequalities* (Springer, New York, 2009)
3. G. Anastassiou, *Integer and Fractional Self Adjoint Operator Opial type Inequalities*, J. Comput. Analy. Appl. (2016)
4. R.C. Brown, D.B. Hinton, Opial's inequality and oscillation of 2nd order equations. Proc. AMS **125**(4), 1123–1129 (1997)
5. S.S. Dragomir, *Inequalities for functions of selfadjoint operators on Hilbert Spaces* www.ajmaa.org/RGMIA/monographs/InFuncOp.pdf (2011)
6. S. Dragomir, *Operator Inequalities of Ostrowski and Trapezoidal Type* (Springer, New York, 2012)
7. T. Furuta, J. Mićić Hot, J. Pečaric, Y. Seo, *Mond-Pečaric Method in Operator Inequalities. Inequalities for Bounded Selfadjoint Operators on a Hilbert Space*, Element, Zagreb, 2005
8. G.-S. Yang, On a certain result of Z. Opial. Proc. Jpn. Acad. **42**, 78–83 (1966)
9. G. Helmberg, *Introduction to Spectral Theory in Hilbert Space* (John Wiley & Sons Inc, New York, 1969)
10. Z. Opial, Sur une inégalité. Ann. Polon. Math. **8**, 29–32 (1960)
11. B.G. Pachpatte, *Mathematical Inequalities* (Elsevier, North-Holand Mathematical Library, Amsterdam, Boston, 2005)
12. W.C. Troy, *On the Opial-Olech-Beesack inequalities*, USA-Chile Workshop on Nonlinear Analysis, Electron. J. Diff. Eqns. Conf., 06 (2001), 297–301, http://ejde.math.swt.edu or http://ejde.math.unt.edu

Chapter 7
Self Adjoint Operator Chebyshev-Grüss Inequalities

We present here very general self adjoint operator Chebyshev-Grüss type inequalities to all cases. We give applications. It follows [2].

7.1 Motivation

Here we mention the following inspiring and motivating results.

Theorem 7.1 (Čebyšev, 1882, [3]). *Let $f, g : [a, b] \to \mathbb{R}$ absolutely continuous functions. If $f', g' \in L_\infty([a, b])$, then*

$$\left| \frac{1}{b-a} \int_a^b f(x) g(x) dx - \left(\frac{1}{b-a} \int_a^b f(x) dx \right) \left(\frac{1}{b-a} \int_a^b g(x) dx \right) \right| \tag{7.1}$$

$$\leq \frac{1}{12} (b-a)^2 \|f'\|_\infty \|g'\|_\infty .$$

Also we mention

Theorem 7.2 (Grüss, 1935, [8]). *Let f, g integrable functions from $[a, b]$ into $\mathbb{R}$, such that $m \leq f(x) \leq M$, $\rho \leq g(x) \leq \sigma$, for all $x \in [a, b]$, where $m, M, \rho, \sigma \in \mathbb{R}$. Then*

$$\left| \frac{1}{b-a} \int_a^b f(x) g(x) dx - \left(\frac{1}{b-a} \int_a^b f(x) dx \right) \left(\frac{1}{b-a} \int_a^b g(x) dx \right) \right| \tag{7.2}$$

$$\leq \frac{1}{4} (M-m)(\sigma-\rho).$$

G.A. Anastassiou, *Intelligent Comparisons II: Operator Inequalities and Approximations*, Studies in Computational Intelligence 699,
DOI 10.1007/978-3-319-51475-8_7

A recent result follows:

Theorem 7.3 (Anastassiou, 2011: see [1], pp. 312–313). *Let $f, g : [a, b] \to \mathbb{R}$, $n \in \mathbb{N}$, $f^{(n-1)}, g^{(n-1)}$ are absolutely continuous on $[a, b]$. Denote*

$$F_{n-1}^{f}(x) := \sum_{k=1}^{n-1} \left(\frac{n-k}{k!}\right) \left(\frac{f^{(k-1)}(b)(x-b)^k - f^{(k-1)}(a)(x-a)^k}{b-a}\right), \tag{7.3}$$

$\left(F_0^f(x) = 0\right)$,

$$F_{n-1}^{g}(x) := \sum_{k=1}^{n-1} \left(\frac{n-k}{k!}\right) \left(\frac{g^{(k-1)}(b)(x-b)^k - g^{(k-1)}(a)(x-a)^k}{b-a}\right) \tag{7.4}$$

$\left(F_0^g(x) = 0\right)$, *and*

$$\Delta_{(f,g)} := \int_a^b f(x) g(x) dx - \frac{n}{b-a} \left(\int_a^b f(x) dx\right) \left(\int_a^b g(x) dx\right) \tag{7.5}$$

$$-\frac{1}{2}\left[\int_a^b \left(g(x) F_{n-1}^f(x) + f(x) F_{n-1}^g(x)\right) dx\right].$$

(1) *If $f^{(n)}, g^{(n)} \in L_\infty([a, b])$, then*

$$\left|\Delta_{(f,g)}\right| \le \frac{(b-a)^{n+1}}{(n+2)!} \left[\|f\|_\infty \left\|g^{(n)}\right\|_\infty + \|g\|_\infty \left\|f^{(n)}\right\|_\infty\right]. \tag{7.6}$$

(2) *If $f^{(n)}, g^{(n)} \in L_p([a, b])$, where $p, q > 1$ such that $\frac{1}{p} + \frac{1}{q} = 1$, then*

$$\left|\Delta_{(f,g)}\right| \le 2^{-\frac{1}{p}} (qn+2)^{-\frac{1}{q}} (B(q(n-1)+1, q+1))^{\frac{1}{q}} \frac{(b-a)^{n-1+\frac{2}{q}}}{(n-1)!} \tag{7.7}$$

$$\times \left[\|f\|_p \left\|g^{(n)}\right\|_p + \|g\|_p \left\|f^{(n)}\right\|_p\right].$$

When $p = q = 2$, it holds

$$\left|\Delta_{(f,g)}\right| \le \frac{(b-a)^n}{(n-1)! 2\sqrt{n(n+1)\left(4n^2-1\right)}} \left[\|f\|_2 \left\|g^{(n)}\right\|_2 + \|g\|_2 \left\|f^{(n)}\right\|_2\right]. \tag{7.8}$$

(3) *With respect to* $\|\cdot\|_1$ *it holds*

$$\left|\Delta_{(f,g)}\right| \leq \frac{(b-a)^n}{2\,(n+1)!}\left(\|f\|_1 \left\|g^{(n)}\right\|_\infty + \|g\|_1 \left\|f^{(n)}\right\|_\infty\right). \tag{7.9}$$

7.2 Background

Let A be a selfadjoint linear operator on a complex Hilbert space $(H; \langle\cdot,\cdot\rangle)$. The Gelfand map establishes a $*-$isometrically isomorphism Φ between the set $C\,(Sp\,(A))$ of all continuous functions defined on the spectrum of A, denoted $Sp\,(A)$, and the C^*-algebra $C^*\,(A)$ generated by A and the identity operator 1_H on H as follows (see e.g. [7, p. 3]):

For any $f, g \in C\,(Sp\,(A))$ and any $\alpha, \beta \in \mathbb{C}$ we have

(i) $\Phi\,(\alpha f + \beta g) = \alpha\Phi\,(f) + \beta\Phi(g)$;
(ii) $\Phi\,(fg) = \Phi\,(f)\,\Phi\,(g)$ (the operation composition is on the right) and $\Phi\left(\overline{f}\right) = (\Phi(f))^*$;
(iii) $\|\Phi\,(f)\| = \|f\| := \sup_{t\in Sp(A)} |f(t)|$;
(iv) $\Phi\,(f_0) = 1_H$ and $\Phi\,(f_1) = A$, where $f_0\,(t) = 1$ and $f_1\,(t) = t$, for $t \in Sp(A)$.

With this notation we define

$$f\,(A) := \Phi\,(f), \quad \text{for all } f \in C\,(Sp\,(A)),$$

and we call it the continuous functional calculus for a selfadjoint operator A.

If A is a selfadjoint operator and f is a real valued continuous function on $Sp\,(A)$ then $f\,(t) \geq 0$ for any $t \in Sp\,(A)$ implies that $f\,(A) \geq 0$, i.e. $f\,(A)$ is a positive operator on H. Moreover, if both f and g are real valued functions on $Sp\,(A)$ then the following important property holds:

(P) $f\,(t) \geq g\,(t)$ for any $t \in Sp\,(A)$, implies that $f\,(A) \geq g\,(A)$ in the operator order of $B\,(H)$.

Equivalently, we use (see [5], pp. 7–8):

Let U be a selfadjoint operator on the complex Hilbert space $(H, \langle\cdot,\cdot\rangle)$ with the spectrum $Sp\,(U)$ included in the interval $[m, M]$ for some real numbers $m < M$ and $\{E_\lambda\}_\lambda$ be its spectral family.

Then for any continuous function $f : [m, M] \to \mathbb{C}$, it is well known that we have the following spectral representation in terms of the Riemann-Stieljes integral:

$$\langle f\,(U)\,x, y\rangle = \int_{m-0}^{M} f\,(\lambda)\,d(\langle E_\lambda x, y\rangle), \tag{7.10}$$

for any $x, y \in H$. The function $g_{x,y}(\lambda) := \langle E_\lambda x, y\rangle$ is of bounded variation on the interval $[m, M]$, and

$$g_{x,y}(m-0) = 0 \quad \text{and} \quad g_{x,y}(M) = \langle x, y\rangle,$$

for any $x, y \in H$. Furthermore, it is known that $g_x(\lambda) := \langle E_\lambda x, x\rangle$ is increasing and right continuous on $[m, M]$.

In this chapter we will be using a lot the formula

$$\langle f(U)x, x\rangle = \int_{m-0}^{M} f(\lambda)\, d(\langle E_\lambda x, x\rangle), \ \forall x \in H. \tag{7.11}$$

As a symbol we can write

$$f(U) = \int_{m-0}^{M} f(\lambda)\, dE_\lambda. \tag{7.12}$$

Above, $m = \min\{\lambda | \lambda \in Sp(U)\} := \min Sp(U)$, $M = \max\{\lambda | \lambda \in Sp(U)\} := \max Sp(U)$. The projections $\{E_\lambda\}_{\lambda \in \mathbb{R}}$, are called the spectral family of A, with the properties:

(a) $E_\lambda \leq E_{\lambda'}$ for $\lambda \leq \lambda'$;
(b) $E_{m-0} = 0_H$ (zero operator), $E_M = 1_H$ (identity operator) and $E_{\lambda+0} = E_\lambda$ for all $\lambda \in \mathbb{R}$.

Furthermore

$$E_\lambda := \varphi_\lambda(U), \ \forall \lambda \in \mathbb{R}, \tag{7.13}$$

is a projection which reduces U, with

$$\varphi_\lambda(s) := \begin{cases} 1, & \text{for } -\infty < s \leq \lambda, \\ 0, & \text{for } \lambda < s < +\infty. \end{cases}$$

The spectral family $\{E_\lambda\}_{\lambda \in \mathbb{R}}$ determines uniquely the self-adjoint operator U and vice versa.

For more on the topic see [9], pp. 256–266, and for more details see there pp. 157–266. See also [4].

Some more basics are given (we follow [5], pp. 1–5):

Let $(H; \langle \cdot, \cdot \rangle)$ be a Hilbert space over $\mathbb{C}$. A bounded linear operator A defined on H is selfjoint, i.e., $A = A^*$, iff $\langle Ax, x\rangle \in \mathbb{R}, \forall x \in H$, and if A is selfadjoint, then

$$\|A\| = \sup_{x \in H : \|x\| = 1} |\langle Ax, x\rangle|. \tag{7.14}$$

Let A, B be selfadjoint operators on H. Then $A \leq B$ iff $\langle Ax, x\rangle \leq \langle Bx, x\rangle$, $\forall x \in H$.

In particular, A is called positive if $A \geq 0$.

Denote by

$$\mathcal{P} := \left\{ \varphi(s) := \sum_{k=0}^{n} \alpha_k s^k | n \geq 0, \alpha_k \in \mathbb{C}, 0 \leq k \leq n \right\}. \tag{7.15}$$

If $A \in \mathcal{B}(H)$ (the Banach algebra of all bounded linear operators defined on H, i.e. from H into itself) is selfadjoint, and $\varphi(s) \in \mathcal{P}$ has real coefficients, then $\varphi(A)$ is selfadjoint, and

$$\|\varphi(A)\| = \max\{|\varphi(\lambda)|, \lambda \in Sp(A)\}. \tag{7.16}$$

If φ is any function defined on $\mathbb{R}$ we define

$$\|\varphi\|_A := \sup\{|\varphi(\lambda)|, \lambda \in Sp(A)\}. \tag{7.17}$$

If A is selfadjoint operator on Hilbert space H and φ is continuous and given that $\varphi(A)$ is selfadjoint, then $\|\varphi(A)\| = \|\varphi\|_A$. And if φ is a continuous real valued function so it is $|\varphi|$, then $\varphi(A)$ and $|\varphi|(A) = |\varphi(A)|$ are selfadjoint operators (by [5], p. 4, Theorem 7).

Hence it holds

$$\||\varphi(A)|\| = \||\varphi|\|_A = \sup\{||\varphi(\lambda)||, \lambda \in Sp(A)\}$$

$$= \sup\{|\varphi(\lambda)|, \lambda \in Sp(A)\} = \|\varphi\|_A = \|\varphi(A)\|,$$

that is

$$\||\varphi(A)|\| = \|\varphi(A)\|. \tag{7.18}$$

For a selfadjoint operator $A \in \mathcal{B}(H)$ which is positive, there exists a unique positive selfadjoint operator $B := \sqrt{A} \in \mathcal{B}(H)$ such that $B^2 = A$, that is $\left(\sqrt{A}\right)^2 = A$. We call B the square root of A.

Let $A \in \mathcal{B}(H)$, then A^*A is selfadjoint and positive. Define the "operator absolute value" $|A| := \sqrt{A^*A}$. If $A = A^*$, then $|A| = \sqrt{A^2}$.

For a continuous real valued function φ we observe the following:

$$|\varphi(A)| \text{ (the functional absolute value) } = \int_{m-0}^{M} |\varphi(\lambda)| \, dE_\lambda =$$

$$\int_{m-0}^{M} \sqrt{(\varphi(\lambda))^2} dE_\lambda = \sqrt{(\varphi(A))^2} = |\varphi(A)| \text{ (operator absolute value)},$$

where A is a selfadjoint operator.

That is we have

$$|\varphi(A)| \text{ (functional absolute value) } = |\varphi(A)| \text{ (operator absolute value).} \quad (7.19)$$

Let $A, B \in \mathcal{B}(H)$, then

$$\|AB\| \leq \|A\| \, \|B\|, \quad (7.20)$$

by Banach algebra property.

7.3 Main Results

Next we present very general Chebyshev-Grüss type operator inequalities based on Fink's ([6]) identity.

Then we specialize them for $n = 1$.

We give

Theorem 7.4 *Let $n \in \mathbb{N}$ and $f, g \in C^n([a, b])$ with $[m, M] \subset (a, b)$, $m < M$. Here A is a selfadjoint linear operator on the Hilbert space H with spectrum $Sp(A) \subseteq [m, M]$. We consider any $x \in H : \|x\| = 1$.*

Then

$$\langle (\Delta(f, g))(A)x, x\rangle :=$$

$$\Bigg| \langle f(A) g(A) x, x\rangle - \langle f(A)x, x\rangle \cdot \langle g(A)x, x\rangle - \frac{1}{2(M-m)} \sum_{k=1}^{n-1} \left(\frac{n-k}{k!}\right) \cdot$$

$$\{\{g^{(k-1)}(m)\left[\langle f(A)x, x\rangle \left\langle (A - m1_H)^k x, x\right\rangle - \left\langle f(A)(A - m1_H)^k x, x\right\rangle\right] +$$

$$g^{(k-1)}(M)\left[\left\langle f(A)(A - M1_H)^k x, x\right\rangle - \langle f(A)x, x\rangle \left\langle (A - M1_H)^k x, x\right\rangle\right]\} +$$

$$\{f^{(k-1)}(m)\left[\langle g(A)x, x\rangle \left\langle (A - m1_H)^k x, x\right\rangle - \left\langle g(A)(A - m1_H)^k x, x\right\rangle\right] +$$

$$f^{(k-1)}(M)\left[\left\langle g(A)(A - M1_H)^k x, x\right\rangle - \langle g(A)x, x\rangle \left\langle (A - M1_H)^k x, x\right\rangle\right]\}\}\Bigg| \leq \quad (7.21)$$

$$\frac{1}{(n+1)!(M-m)} \left[\left\|g^{(n)}\right\|_{\infty,[m,M]} \|f(A)\| + \left\|f^{(n)}\right\|_{\infty,[m,M]} \|g(A)\|\right] \cdot$$

$$\left[\left\|(M1_H - A)^{n+1}\right\| + \left\|(A - m1_H)^{n+1}\right\|\right].$$

Proof Let $a, b \in \mathbb{R}$; $f, g : [a, b] \to \mathbb{R}$, $n \in \mathbb{N}$, where $f^{(n)}, g^{(n)}$ are continuous on $[a, b]$. Then by Fink ([6]) we have

$$f(\lambda) = \frac{n}{b-a} \int_a^b f(t)\, dt -$$

$$\sum_{k=1}^{n-1} \left(\frac{n-k}{k!} \right) \left(\frac{f^{(k-1)}(a)(\lambda - a)^k - f^{(k-1)}(b)(\lambda - b)^k}{b-a} \right) \tag{7.22}$$

$$+ \frac{1}{(n-1)!(b-a)} \int_a^b (\lambda - t)^{n-1} k^*(t, \lambda) f^{(n)}(t)\, dt,$$

where

$$k^*(t, \lambda) := \begin{cases} t - a, & a \le t \le \lambda \le b, \\ t - b, & a \le \lambda < t \le b, \end{cases} \quad \forall \lambda \in [a, b].$$

When $n = 1$ the sum $\sum_{k=1}^{n-1}$ is (7.22) is zero.
Similarly, we get

$$g(\lambda) = \frac{n}{b-a} \int_a^b g(t)\, dt -$$

$$\sum_{k=1}^{n-1} \left(\frac{n-k}{k!} \right) \left(\frac{g^{(k-1)}(a)(\lambda - a)^k - g^{(k-1)}(b)(\lambda - b)^k}{b-a} \right) \tag{7.23}$$

$$+ \frac{1}{(n-1)!(b-a)} \int_a^b (\lambda - t)^{n-1} k^*(t, \lambda) g^{(n)}(t)\, dt, \quad \forall \lambda \in [a, b].$$

Here A is a selfadjoint operator on the Hilbert space H with the spectrum $Sp(A) \subseteq [m, M]$ for some real numbers $m < M$, $\{E_\lambda\}_\lambda$ is the spectral family of A, and $[m, M] \subset (a, b)$.

Therefore we have

$$f(\lambda) = \frac{n}{M-m} \int_m^M f(t)\, dt -$$

$$\sum_{k=1}^{n-1} \left(\frac{n-k}{k!} \right) \left(\frac{f^{(k-1)}(m)(\lambda - m)^k - f^{(k-1)}(M)(\lambda - M)^k}{M-m} \right) \tag{7.24}$$

$$+ \frac{1}{(n-1)!(M-m)} \int_m^M (\lambda - t)^{n-1} k(t, \lambda) f^{(n)}(t)\, dt,$$

where

$$k(t,\lambda) := \begin{cases} t-m, & m \le t \le \lambda \le M, \\ t-M, & m \le \lambda < t \le M, \end{cases} \tag{7.25}$$

and

$$g(\lambda) = \frac{n}{M-m}\int_m^M g(t)\,dt -$$

$$\sum_{k=1}^{n-1}\left(\frac{n-k}{k!}\right)\left(\frac{g^{(k-1)}(m)(\lambda-m)^k - g^{(k-1)}(M)(\lambda-M)^k}{M-m}\right) \tag{7.26}$$

$$+\frac{1}{(n-1)!(M-m)}\int_m^M (\lambda-t)^{n-1}k(t,\lambda)\,g^{(n)}(t)\,dt, \quad \forall\,\lambda \in [m,M].$$

By applying the spectral representation theorem on (7.24) and (7.26), i.e. integrating against E_λ over $[m, M]$, see (7.12), we obtain:

$$f(A) = \left(\frac{n}{M-m}\int_m^M f(t)\,dt\right)1_H - \sum_{k=1}^{n-1}\left(\frac{n-k}{k!}\right)\cdot$$

$$\left(\frac{f^{(k-1)}(m)(A-m1_H)^k - f^{(k-1)}(M)(A-M1_H)^k}{M-m}\right)+ \tag{7.27}$$

$$\frac{1}{(n-1)!(M-m)}\int_{m-0}^M\left(\int_m^M (\lambda-t)^{n-1}k(t,\lambda)\,f^{(n)}(t)\,dt\right)dE_\lambda,$$

and

$$g(A) = \left(\frac{n}{M-m}\int_m^M g(t)\,dt\right)1_H - \sum_{k=1}^{n-1}\left(\frac{n-k}{k!}\right)\cdot$$

$$\left(\frac{g^{(k-1)}(m)(A-m1_H)^k - g^{(k-1)}(M)(A-M1_H)^k}{M-m}\right)+ \tag{7.28}$$

$$\frac{1}{(n-1)!(M-m)}\int_{m-0}^M\left(\int_m^M (\lambda-t)^{n-1}k(t,\lambda)\,g^{(n)}(t)\,dt\right)dE_\lambda.$$

We notice that

$$g(A)\,f(A) = f(A)\,g(A), \tag{7.29}$$

to be used next.

Hence it holds

$$g(A) f(A) = \left(\frac{n}{M-m} \int_m^M f(t) dt \right) g(A) - \sum_{k=1}^{n-1} \left(\frac{n-k}{k!} \right) \cdot$$

$$\left(\frac{f^{(k-1)}(m) g(A) (A - m1_H)^k - f^{(k-1)}(M) g(A) (A - M1_H)^k}{M-m} \right) + \quad (7.30)$$

$$\frac{1}{(n-1)!(M-m)} g(A) \int_{m-0}^M \left(\int_m^M (\lambda - t)^{n-1} k(t,\lambda) f^{(n)}(t) dt \right) dE_\lambda,$$

and

$$f(A) g(A) = \left(\frac{n}{M-m} \int_m^M g(t) dt \right) f(A) - \sum_{k=1}^{n-1} \left(\frac{n-k}{k!} \right) \cdot$$

$$\left(\frac{g^{(k-1)}(m) f(A) (A - m1_H)^k - g^{(k-1)}(M) f(A) (A - M1_H)^k}{M-m} \right) + \quad (7.31)$$

$$\frac{1}{(n-1)!(M-m)} f(A) \int_{m-0}^M \left(\int_m^M (\lambda - t)^{n-1} k(t,\lambda) g^{(n)}(t) dt \right) dE_\lambda.$$

Here from on we consider $x \in H : \|x\| = 1$; immediately we get $\int_{m-0}^M d \langle E_\lambda x, x \rangle = 1$.

Then it holds

$$\langle f(A) x, x \rangle = \frac{n}{M-m} \int_m^M f(s) ds - \sum_{k=1}^{n-1} \left(\frac{n-k}{k!} \right) \cdot$$

$$\left(\frac{f^{(k-1)}(m) \langle (A - m1_H)^k x, x \rangle - f^{(k-1)}(M) \langle (A - M1_H)^k x, x \rangle}{M-m} \right) + \quad (7.32)$$

$$\frac{1}{(n-1)!(M-m)} \int_{m-0}^M \left(\int_m^M (\lambda - s)^{n-1} k(s,\lambda) f^{(n)}(s) ds \right) d \langle E_\lambda x, x \rangle,$$

and

$$\langle g(A) x, x \rangle = \frac{n}{M-m} \int_m^M g(s) ds - \sum_{k=1}^{n-1} \left(\frac{n-k}{k!} \right) \cdot$$

$$\left(\frac{g^{(k-1)}(m)\left\langle (A-m1_H)^k x,x\right\rangle - g^{(k-1)}(M)\left\langle (A-M1_H)^k x,x\right\rangle}{M-m}\right)+ \tag{7.33}$$

$$\frac{1}{(n-1)!(M-m)}\int_{m-0}^{M}\left(\int_m^M (\lambda-s)^{n-1}k(s,\lambda)g^{(n)}(s)\,ds\right)d\langle E_\lambda x,x\rangle.$$

Then we get

$$\langle f(A)x,x\rangle\langle g(A)x,x\rangle = \left(\frac{n}{M-m}\int_m^M f(s)\,ds\right)\langle g(A)x,x\rangle - \sum_{k=1}^{n-1}\left(\frac{n-k}{k!}\right)\cdot$$

$$\left(\frac{f^{(k-1)}(m)\langle g(A)x,x\rangle\left\langle (A-m1_H)^k x,x\right\rangle - f^{(k-1)}(M)\langle g(A)x,x\rangle\left\langle (A-M1_H)^k x,x\right\rangle}{M-m}\right) \tag{7.34}$$

$$+\frac{\langle g(A)x,x\rangle}{(n-1)!(M-m)}\int_{m-0}^{M}\left(\int_m^M (\lambda-s)^{n-1}k(s,\lambda)f^{(n)}(s)\,ds\right)d\langle E_\lambda x,x\rangle,$$

and

$$\langle g(A)x,x\rangle\langle f(A)x,x\rangle = \frac{n}{M-m}\left(\int_m^M g(s)\,ds\right)\langle f(A)x,x\rangle - \sum_{k=1}^{n-1}\left(\frac{n-k}{k!}\right)\cdot$$

$$\left(\frac{g^{(k-1)}(m)\langle f(A)x,x\rangle\left\langle (A-m1_H)^k x,x\right\rangle - g^{(k-1)}(M)\langle f(A)x,x\rangle\left\langle (A-M1_H)^k x,x\right\rangle}{M-m}\right) \tag{7.35}$$

$$+\frac{\langle f(A)x,x\rangle}{(n-1)!(M-m)}\int_{m-0}^{M}\left(\int_m^M (\lambda-s)^{n-1}k(s,\lambda)g^{(n)}(s)\,ds\right)d\langle E_\lambda x,x\rangle.$$

Hence we obtain

$$\langle f(A)g(A)x,x\rangle = \left(\frac{n}{M-m}\int_m^M f(s)\,ds\right)\langle g(A)x,x\rangle - \sum_{k=1}^{n-1}\left(\frac{n-k}{k!}\right)\cdot$$

$$\left(\frac{f^{(k-1)}(m)\left\langle g(A)(A-m1_H)^k x,x\right\rangle - f^{(k-1)}(M)\left\langle g(A)(A-M1_H)^k x,x\right\rangle}{M-m}\right)+ \tag{7.36}$$

$$\frac{1}{(n-1)!(M-m)}\left\langle\left(g(A)\int_{m-0}^{M}\left(\int_m^M (\lambda-s)^{n-1}k(s,\lambda)f^{(n)}(s)\,ds\right)dE_\lambda\right)x,x\right\rangle,$$

and

$$\langle f(A) g(A) x, x\rangle = \left(\frac{n}{M-m}\int_m^M g(s)\,ds\right)\langle f(A) x, x\rangle - \sum_{k=1}^{n-1}\left(\frac{n-k}{k!}\right)\cdot$$

$$\left(\frac{g^{(k-1)}(m)\left\langle f(A)(A-m1_H)^k x, x\right\rangle - g^{(k-1)}(M)\left\langle f(A)(A-M1_H)^k x, x\right\rangle}{M-m}\right)+ \tag{7.37}$$

$$\frac{1}{(n-1)!(M-m)}\left\langle\left(f(A)\int_{m-0}^M\left(\int_m^M(\lambda-s)^{n-1}k(s,\lambda)g^{(n)}(s)\,ds\right)dE_\lambda\right)x, x\right\rangle.$$

By (7.37)–(7.35) we obtain

$$E := \langle f(A) g(A) x, x\rangle - \langle f(A) x, x\rangle\langle g(A) x, x\rangle = -\sum_{k=1}^{n-1}\left(\frac{n-k}{k!}\right)\cdot$$

$$\left(\frac{g^{(k-1)}(m)\left\langle f(A)(A-m1_H)^k x, x\right\rangle - g^{(k-1)}(M)\left\langle f(A)(A-M1_H)^k x, x\right\rangle}{M-m}\right)$$

$$+\sum_{k=1}^{n-1}\left(\frac{n-k}{k!}\right)\cdot$$

$$\left(\frac{g^{(k-1)}(m)\langle f(A)x, x\rangle\left\langle (A-m1_H)^k x, x\right\rangle - g^{(k-1)}(M)\langle f(A)x, x\rangle\left\langle (A-M1_H)^k x, x\right\rangle}{M-m}\right)$$

$$+\frac{1}{(n-1)!(M-m)}\left[\left\langle\left(f(A)\int_{m-0}^M\left(\int_m^M(\lambda-s)^{n-1}k(s,\lambda)g^{(n)}(s)\,ds\right)dE_\lambda\right)x, x\right\rangle\right. \tag{7.38}$$

$$\left.-\langle f(A)x, x\rangle\int_{m-0}^M\left(\int_m^M(\lambda-s)^{n-1}k(s,\lambda)g^{(n)}(s)\,ds\right)d\langle E_\lambda x, x\rangle\right].$$

By (7.36)–(7.34) we also get

$$E := \langle f(A) g(A) x, x\rangle - \langle f(A) x, x\rangle\langle g(A) x, x\rangle = -\sum_{k=1}^{n-1}\left(\frac{n-k}{k!}\right)\cdot$$

$$\left(\frac{f^{(k-1)}(m)\left\langle g(A)(A-m1_H)^k x, x\right\rangle - f^{(k-1)}(M)\left\langle g(A)(A-M1_H)^k x, x\right\rangle}{M-m}\right)$$

$$+\sum_{k=1}^{n-1}\left(\frac{n-k}{k!}\right)\cdot$$

$$\left(\frac{f^{(k-1)}(m)\langle g(A)x,x\rangle\left\langle (A-m1_H)^k x,x\right\rangle - f^{(k-1)}(M)\langle g(A)x,x\rangle\left\langle (A-M1_H)^k x,x\right\rangle}{M-m}\right)$$

$$+\frac{1}{(n-1)!(M-m)}\left[\left\langle\left(g(A)\int_{m-0}^{M}\left(\int_m^M(\lambda-s)^{n-1}k(s,\lambda)f^{(n)}(s)\,ds\right)dE_\lambda\right)x,x\right\rangle\right. \tag{7.39}$$

$$\left.-\langle g(A)x,x\rangle\int_{m-0}^{M}\left(\int_m^M(\lambda-s)^{n-1}k(s,\lambda)f^{(n)}(s)\,ds\right)d\langle E_\lambda x,x\rangle\right].$$

Consequently, we get that

$$2E=\frac{1}{M-m}\sum_{k=1}^{n-1}\left(\frac{n-k}{k!}\right)\cdot$$

$$\left\{\left\{g^{(k-1)}(m)\left[\langle f(A)x,x\rangle\left\langle (A-m1_H)^k x,x\right\rangle-\left\langle f(A)(A-m1_H)^k x,x\right\rangle\right]+\right.\right.$$

$$\left.g^{(k-1)}(M)\left[\left\langle f(A)(A-M1_H)^k x,x\right\rangle-\langle f(A)x,x\rangle\left\langle (A-M1_H)^k x,x\right\rangle\right]\right\}+$$

$$\left\{f^{(k-1)}(m)\left[\langle g(A)x,x\rangle\left\langle (A-m1_H)^k x,x\right\rangle-\left\langle g(A)(A-m1_H)^k x,x\right\rangle\right]+\right.$$

$$\left.\left.f^{(k-1)}(M)\left[\left\langle g(A)(A-M1_H)^k x,x\right\rangle-\langle g(A)x,x\rangle\left\langle (A-M1_H)^k x,x\right\rangle\right]\right\}\right\}$$

$$+\frac{1}{(n-1)!(M-m)}\cdot$$

$$\left\{\left[\left\langle\left(f(A)\int_{m-0}^{M}\left(\int_m^M(\lambda-s)^{n-1}k(s,\lambda)g^{(n)}(s)\,ds\right)dE_\lambda\right)x,x\right\rangle-\right.\right. \tag{7.40}$$

$$\left.\langle f(A)x,x\rangle\int_{m-0}^{M}\left(\int_m^M(\lambda-s)^{n-1}k(s,\lambda)g^{(n)}(s)\,ds\right)d\langle E_\lambda x,x\rangle\right]+$$

$$\left[\left\langle\left(g(A)\int_{m-0}^{M}\left(\int_m^M(\lambda-s)^{n-1}k(s,\lambda)f^{(n)}(s)\,ds\right)dE_\lambda\right)x,x\right\rangle-\right.$$

$$\left.\left.\langle g(A)x,x\rangle\int_{m-0}^{M}\left(\int_m^M(\lambda-s)^{n-1}k(s,\lambda)f^{(n)}(s)\,ds\right)d\langle E_\lambda x,x\rangle\right]\right\}.$$

We find

$$\langle f(A)g(A)x,x\rangle-\langle f(A)x,x\rangle\langle g(A)x,x\rangle-\frac{1}{2(M-m)}\sum_{k=1}^{n-1}\left(\frac{n-k}{k!}\right)\cdot$$

$$\{\{g^{(k-1)}(m)\left[\langle f(A)x,x\rangle\left\langle (A-m1_H)^k x,x\right\rangle - \left\langle f(A)(A-m1_H)^k x,x\right\rangle\right]+$$

$$g^{(k-1)}(M)\left[\left\langle f(A)(A-M1_H)^k x,x\right\rangle - \langle f(A)x,x\rangle\left\langle (A-M1_H)^k x,x\right\rangle\right]\}+$$

$$\{f^{(k-1)}(m)\left[\langle g(A)x,x\rangle\left\langle (A-m1_H)^k x,x\right\rangle - \left\langle g(A)(A-m1_H)^k x,x\right\rangle\right]+$$

$$f^{(k-1)}(M)\left[\left\langle g(A)(A-M1_H)^k x,x\right\rangle - \langle g(A)x,x\rangle\left\langle (A-M1_H)^k x,x\right\rangle\right]\}\}$$

$$= \frac{1}{2(n-1)!(M-m)}\cdot$$

$$\left\{\left[\left\langle\left(f(A)\int_{m-0}^{M}\left(\int_m^M (\lambda-s)^{n-1}k(s,\lambda)g^{(n)}(s)\,ds\right)dE_\lambda\right)x,x\right\rangle - \right.\right. \tag{7.41}$$

$$\left.\langle f(A)x,x\rangle\int_{m-0}^{M}\left(\int_m^M (\lambda-s)^{n-1}k(s,\lambda)g^{(n)}(s)\,ds\right)d\langle E_\lambda x,x\rangle\right]+$$

$$\left[\left\langle\left(g(A)\int_{m-0}^{M}\left(\int_m^M (\lambda-s)^{n-1}k(s,\lambda)f^{(n)}(s)\,ds\right)dE_\lambda\right)x,x\right\rangle - \right. \tag{7.42}$$

$$\left.\left.\langle g(A)x,x\rangle\int_{m-0}^{M}\left(\int_m^M (\lambda-s)^{n-1}k(s,\lambda)f^{(n)}(s)\,ds\right)d\langle E_\lambda x,x\rangle\right]\right\} =: R.$$

Hence we have

$$|R| \le \frac{1}{2(n-1)!(M-m)}\cdot$$

$$\left\{\left[\left|\left\langle\left(f(A)\int_{m-0}^{M}\left(\int_m^M (\lambda-s)^{n-1}k(s,\lambda)g^{(n)}(s)\,ds\right)dE_\lambda\right)x,x\right\rangle\right|\right.\right.$$

$$\left.+\left|\langle f(A)x,x\rangle\right|\left|\int_{m-0}^{M}\left(\int_m^M (\lambda-s)^{n-1}k(s,\lambda)g^{(n)}(s)\,ds\right)d\langle E_\lambda x,x\rangle\right|\right]+ \tag{7.43}$$

$$\left[\left|\left\langle\left(g(A)\int_{m-0}^{M}\left(\int_m^M (\lambda-s)^{n-1}k(s,\lambda)f^{(n)}(s)\,ds\right)dE_\lambda\right)x,x\right\rangle\right|+\right.$$

$$\left.\left.\left|\langle g(A)x,x\rangle\right|\left|\int_{m-0}^{M}\left(\int_m^M (\lambda-s)^{n-1}k(s,\lambda)f^{(n)}(s)\,ds\right)d\langle E_\lambda x,x\rangle\right|\right]\right\}$$

(here notice that

$$\left|\int_m^M (\lambda - s)^{n-1} k(s,\lambda) g^{(n)}(s)\, ds\right| \leq \frac{\left\|g^{(n)}\right\|_{\infty,[m,M]}}{n(n+1)} \left[(M-\lambda)^{n+1} + (\lambda - m)^{n+1}\right])$$

$$\leq \frac{1}{2(n-1)!(M-m)} \left\{ \left\| f(A) \int_{m-0}^M \left(\int_m^M (\lambda - s)^{n-1} k(s,\lambda) g^{(n)}(s)\, ds \right) dE_\lambda \right\| \right.$$

$$+ \|f(A)\| \frac{\left\|g^{(n)}\right\|_{\infty,[m,M]}}{n(n+1)} \left\{ \left\langle (M1_H - A)^{n+1} x, x \right\rangle + \left\langle (A - m1_H)^{n+1} x, x \right\rangle \right\} +$$

$$\left\| g(A) \int_{m-0}^M \left(\int_m^M (\lambda - s)^{n-1} k(s,\lambda) f^{(n)}(s)\, ds \right) dE_\lambda \right\| + \tag{7.44}$$

$$\left. \|g(A)\| \frac{\left\|f^{(n)}\right\|_{\infty,[m,M]}}{n(n+1)} \left\{ \left\langle (M1_H - A)^{n+1} x, x \right\rangle + \left\langle (A - m1_H)^{n+1} x, x \right\rangle \right\} \right\} \leq$$

$$\frac{1}{2(n-1)!(M-m)} \left\{ \|f(A)\| \left[\left\| \int_{m-0}^M \left(\int_m^M (\lambda - s)^{n-1} k(s,\lambda) g^{(n)}(s)\, ds \right) dE_\lambda \right\| \right. \right.$$

$$\left. + \frac{\left\|g^{(n)}\right\|_{\infty,[m,M]}}{n(n+1)} \left\{ \left\langle (M1_H - A)^{n+1} x, x \right\rangle + \left\langle (A - m1_H)^{n+1} x, x \right\rangle \right\} \right] + \tag{7.45}$$

$$\|g(A)\| \left[\left\| \int_{m-0}^M \left(\int_m^M (\lambda - s)^{n-1} k(s,\lambda) f^{(n)}(s)\, ds \right) dE_\lambda \right\| + \right.$$

$$\left. \left. \frac{\left\|f^{(n)}\right\|_{\infty,[m,M]}}{n(n+1)} \left\{ \left\langle (M1_H - A)^{n+1} x, x \right\rangle + \left\langle (A - m1_H)^{n+1} x, x \right\rangle \right\} \right] \right\} =: (\xi). \tag{7.46}$$

Notice here that

$$\left\| \int_{m-0}^M \left(\int_m^M (\lambda - s)^{n-1} k(s,\lambda) g^{(n)}(s)\, ds \right) dE_\lambda \right\| =$$

$$\sup_{\|x\|=1} \left| \int_{m-0}^M \left(\int_m^M (\lambda - s)^{n-1} k(s,\lambda) g^{(n)}(s)\, ds \right) d\left\langle E_\lambda x, x \right\rangle \right| \leq$$

$$\frac{\left\|g^{(n)}\right\|_{\infty,[m,M]}}{n(n+1)} \sup_{\|x\|=1} \left[\left\langle (M1_H - A)^{n+1} x, x \right\rangle + \left\langle (A - m1_H)^{n+1} x, x \right\rangle \right] \leq \tag{7.47}$$

$$\frac{\left\|g^{(n)}\right\|_{\infty,[m,M]}}{n(n+1)}\left\{\sup_{\|x\|=1}\left\langle (M1_H - A)^{n+1}x, x\right\rangle + \sup_{\|x\|=1}\left\langle (A - m1_H)^{n+1}x, x\right\rangle\right\} \leq$$

$$\frac{\left\|g^{(n)}\right\|_{\infty,[m,M]}}{n(n+1)}\left[\left\|(M1_H - A)^{n+1}\right\| + \left\|(A - m1_H)^{n+1}\right\|\right]. \quad (7.48)$$

We have proved that

$$\left\|\int_{m-0}^{M}\left(\int_{m}^{M}(\lambda - s)^{n-1}k(s,\lambda)\,g^{(n)}(s)\,ds\right)dE_\lambda\right\| \leq$$

$$\frac{\left\|g^{(n)}\right\|_{\infty,[m,M]}}{n(n+1)}\left[\left\|(M1_H - A)^{n+1}\right\| + \left\|(A - m1_H)^{n+1}\right\|\right]. \quad (7.49)$$

A similar estimate to (7.49) holds for $f^{(n)}$.

Hence we obtain by (7.46) and (7.49) that

$$(\xi) \leq \frac{1}{2(n-1)!(M-m)}\left\{\frac{\left\|g^{(n)}\right\|_{\infty,[m,M]}}{n(n+1)}\|f(A)\|\cdot\right.$$

$$\left\{\left[\left\|(M1_H - A)^{n+1}\right\| + \left\|(A - m1_H)^{n+1}\right\|\right]+\right.$$

$$\left.\left[\left\|(M1_H - A)^{n+1}\right\| + \left\|(A - m1_H)^{n+1}\right\|\right]\right\}+ \quad (7.50)$$

$$\left\{\frac{\left\|f^{(n)}\right\|_{\infty,[m,M]}}{n(n+1)}\|g(A)\|\left\{\left[\left\|(M1_H - A)^{n+1}\right\| + \left\|(A - m1_H)^{n+1}\right\|\right]+\right.\right.$$

$$\left.\left.\left[\left\|(M1_H - A)^{n+1}\right\| + \left\|(A - m1_H)^{n+1}\right\|\right]\right\}\right\} =$$

$$\frac{1}{(n+1)!(M-m)}\left\{\left\|g^{(n)}\right\|_{\infty,[m,M]}\|f(A)\|\cdot\right.$$

$$\left[\left\|(M1_H - A)^{n+1}\right\| + \left\|(A - m1_H)^{n+1}\right\|\right]+$$

$$\left.\left\|f^{(n)}\right\|_{\infty,[m,M]}\|g(A)\|\left[\left\|(M1_H - A)^{n+1}\right\| + \left\|(A - m1_H)^{n+1}\right\|\right]\right\} = \quad (7.51)$$

$$\frac{1}{(n+1)!(M-m)}\left[\left\|g^{(n)}\right\|_{\infty,[m,M]}\|f(A)\| + \left\|f^{(n)}\right\|_{\infty,[m,M]}\|g(A)\|\right]\cdot \quad (7.52)$$

$$\left[\left\|(M1_H - A)^{n+1}\right\| + \left\|(A - m1_H)^{n+1}\right\|\right].$$

We have proved that

$$|R| \leq \frac{1}{(n+1)!(M-m)} \left[\left\| g^{(n)} \right\|_{\infty,[m,M]} \|f(A)\| + \left\| f^{(n)} \right\|_{\infty,[m,M]} \|g(A)\| \right] \cdot$$

$$\left[\left\| (M1_H - A)^{n+1} \right\| + \left\| (A - m1_H)^{n+1} \right\| \right], \tag{7.53}$$

that is proving the claim. ∎

We give

Corollary 7.5 ($n = 1$ case of Theorem 7.4) *For every $x \in H : \|x\| = 1$, we obtain that*

$$|\langle f(A) g(A) x, x \rangle - \langle f(A) x, x \rangle \langle g(A) x, x \rangle| \leq$$

$$\frac{1}{2(M-m)} \left[\left\| g' \right\|_{\infty,[m,M]} \|f(A)\| + \left\| f' \right\|_{\infty,[m,M]} \|g(A)\| \right] \cdot$$

$$\left[\left\| (M1_H - A)^2 \right\| + \left\| (A - m1_H)^2 \right\| \right]. \tag{7.54}$$

We present

Theorem 7.6 *All as in Theorem 7.4. Let $p, q > 1 : \frac{1}{p} + \frac{1}{q} = 1$. Then*

$$\langle (\Delta(f,g))(A) x, x \rangle \leq \frac{1}{(n-1)!(M-m)} \left(\frac{\Gamma(p+1)\Gamma(p(n-1)+1)}{\Gamma(pn+2)} \right)^{\frac{1}{p}} \cdot$$

$$\left[\left\| g^{(n)} \right\|_{q,[m,M]} \|f(A)\| + \left\| f^{(n)} \right\|_{q,[m,M]} \|g(A)\| \right] \cdot$$

$$\left[\left\| (M1_H - A)^{n+\frac{1}{p}} \right\| + \left\| (A - m1_H)^{n+\frac{1}{p}} \right\| \right], \tag{7.55}$$

where Γ is the gamma function.

Proof We observe that

$$\left| \int_m^M (\lambda - s)^{n-1} k(s, \lambda) g^{(n)}(s) ds \right| \leq$$

$$\left| \int_m^\lambda (\lambda - s)^{n-1} (s - m) g^{(n)}(s) ds \right| + \left| \int_\lambda^M (\lambda - s)^{n-1} (s - M) g^{(n)}(s) ds \right| \leq \tag{7.56}$$

$$\int_m^\lambda (\lambda - s)^{n-1} (s - m) \left| g^{(n)}(s) \right| ds + \int_\lambda^M (M - s)(s - \lambda)^{n-1} \left| g^{(n)}(s) \right| ds \leq$$

$$\left(\int_m^\lambda \left((\lambda - s)^{n-1}(s-m)\right)^p ds\right)^{\frac{1}{p}} \left\|g^{(n)}\right\|_{q,[m,M]} +$$

$$\left(\int_\lambda^M \left((M - s)(s-\lambda)^{n-1}\right)^p ds\right)^{\frac{1}{p}} \left\|g^{(n)}\right\|_{q,[m,M]} =$$

$$\left\|g^{(n)}\right\|_{q,[m,M]} \left[\left(\int_m^\lambda (\lambda - s)^{(p(n-1)+1)-1}(s-m)^{(p+1)-1} ds\right)^{\frac{1}{p}} + \right. \tag{7.57}$$

$$\left.\left(\int_\lambda^M (M - s)^{(p+1)-1}(s-\lambda)^{(p(n-1)+1)-1} ds\right)^{\frac{1}{p}}\right] =$$

$$\left\|g^{(n)}\right\|_{q,[m,M]} \left(\frac{\Gamma(p+1)\,\Gamma(p(n-1)+1)}{\Gamma(pn+2)}\right)^{\frac{1}{p}} \cdot$$

$$\left[(M-\lambda)^{n+\frac{1}{p}} + (\lambda - m)^{n+\frac{1}{p}}\right], \ \forall\, \lambda \in [m, M].$$

Hence it holds

$$\left|\int_{m-0}^M \left(\int_m^M (\lambda - s)^{n-1} k(s,\lambda)\, g^{(n)}(s)\, ds\right) d\langle E_\lambda x, x\rangle\right| \leq$$

$$\left(\frac{\Gamma(p+1)\,\Gamma(p(n-1)+1)}{\Gamma(pn+2)}\right)^{\frac{1}{p}} \left\|g^{(n)}\right\|_{q,[m,M]} \cdot \tag{7.58}$$

$$\left[\left\|(M1_H - A)^{n+\frac{1}{p}}\right\| + \left\|(A - m1_H)^{n+\frac{1}{p}}\right\|\right].$$

Similarly we obtain

$$\left|\int_{m-0}^M \left(\int_m^M (\lambda - s)^{n-1} k(s,\lambda)\, f^{(n)}(s)\, ds\right) d\langle E_\lambda x, x\rangle\right| \leq$$

$$\left(\frac{\Gamma(p+1)\,\Gamma(p(n-1)+1)}{\Gamma(pn+2)}\right)^{\frac{1}{p}} \left\|f^{(n)}\right\|_{q,[m,M]} \cdot \tag{7.59}$$

$$\left[\left\|(M1_H - A)^{n+\frac{1}{p}}\right\| + \left\|(A - m1_H)^{n+\frac{1}{p}}\right\|\right].$$

We also have

$$\left\| \int_{m-0}^{M} \left(\int_{m}^{M} (\lambda - s)^{n-1} k(s, \lambda) g^{(n)}(s) \, ds \right) dE_{\lambda} \right\| \leq$$

$$\left(\frac{\Gamma(p+1) \Gamma(p(n-1)+1)}{\Gamma(pn+2)} \right)^{\frac{1}{p}} \left\| g^{(n)} \right\|_{q,[m,M]} \cdot \tag{7.60}$$

$$\left[\left\| (M1_H - A)^{n+\frac{1}{p}} \right\| + \left\| (A - m1_H)^{n+\frac{1}{p}} \right\| \right].$$

A similar estimate to (7.60) can be derived for $f^{(n)}$.

Acting as in the proof of Theorem 7.4 we find that

$$|R| \leq \frac{1}{(n-1)!\,(M-m)} \left(\frac{\Gamma(p+1) \Gamma(p(n-1)+1)}{\Gamma(pn+2)} \right)^{\frac{1}{p}} \cdot$$

$$\left[\left\| g^{(n)} \right\|_{q,[m,M]} \| f(A) \| + \left\| f^{(n)} \right\|_{q,[m,M]} \| g(A) \| \right] \cdot$$

$$\left[\left\| (M1_H - A)^{n+\frac{1}{p}} \right\| + \left\| (A - m1_H)^{n+\frac{1}{p}} \right\| \right], \tag{7.61}$$

proving the claim. ■

We present

Corollary 7.7 (to Theorem 7.6, $n = 1$) *It holds*

$$|\langle f(A) g(A) x, x \rangle - \langle f(A) x, x \rangle \langle g(A) x, x \rangle| \leq$$

$$\frac{1}{(M-m)(p+1)^{\frac{1}{p}}} \left[\left\| g' \right\|_{q,[m,M]} \| f(A) \| + \left\| f' \right\|_{q,[m,M]} \| g(A) \| \right] \cdot$$

$$\left[\left\| (M1_H - A)^{1+\frac{1}{p}} \right\| + \left\| (A - m1_H)^{1+\frac{1}{p}} \right\| \right]. \tag{7.62}$$

We give

Theorem 7.8 *All as in Theorem 7.4. Then*

$$\langle (\Delta(f, g))(A) x, x \rangle \leq \frac{(M-m)^{n-1}}{(n-1)!} \cdot$$

$$\left[\left\| g^{(n)} \right\|_{1,[m,M]} \| f(A) \| + \left\| f^{(n)} \right\|_{1,[m,M]} \| g(A) \| \right]. \tag{7.63}$$

Proof We observe that

$$\left|\int_m^M (\lambda - s)^{n-1} k(s,\lambda) g^{(n)}(s)\, ds\right| \le$$

$$\int_m^M |\lambda - s|^{n-1} |k(s,\lambda)| \left|g^{(n)}(s)\right| ds \le \tag{7.64}$$

$$(M-m)^n \int_m^M \left|g^{(n)}(s)\right| ds = (M-m)^n \left\|g^{(n)}\right\|_{1,[m,M]}.$$

Hence it holds

$$\left|\int_{m-0}^M \left(\int_m^M (\lambda - s)^{n-1} k(s,\lambda) g^{(n)}(s)\, ds\right) d\langle E_\lambda x, x\rangle\right| \le (M-m)^n \left\|g^{(n)}\right\|_{1,[m,M]}, \tag{7.65}$$

and similarly,

$$\left|\int_{m-0}^M \left(\int_m^M (\lambda - s)^{n-1} k(s,\lambda) f^{(n)}(s)\, ds\right) d\langle E_\lambda x, x\rangle\right| \le (M-m)^n \left\|f^{(n)}\right\|_{1,[m,M]}; \tag{7.66}$$

the last are valid since

$$\int_{m-0}^M d\langle E_\lambda x, x\rangle = 1, \text{ for } x \in H : \|x\| = 1.$$

Similarly, we obtain

$$\left\|\int_{m-0}^M \left(\int_m^M (\lambda - s)^{n-1} k(s,\lambda) g^{(n)}(s)\, ds\right) dE_\lambda\right\| \le (M-m)^n \left\|g^{(n)}\right\|_{1,[m,M]}. \tag{7.67}$$

A similar estimate to (7.67) can be derived for $f^{(n)}$.

Acting as in the proof of Theorem 7.4 we find that

$$|R| \le \frac{(M-m)^{n-1}}{(n-1)!}\left[\left\|g^{(n)}\right\|_{1,[m,M]} \|f(A)\| + \left\|f^{(n)}\right\|_{1,[m,M]} \|g(A)\|\right], \tag{7.68}$$

proving the claim. ■

We finish with

Corollary 7.9 (to Theorem 7.8, $n = 1$) *It holds*

$$|\langle f(A) g(A) x, x\rangle - \langle f(A) x, x\rangle \langle g(A) x, x\rangle| \le$$

$$\left[\left\|g'\right\|_{1,[m,M]} \|f(A)\| + \left\|f'\right\|_{1,[m,M]} \|g(A)\|\right]. \tag{7.69}$$

References

1. G.A. Anastassiou, *Advances Inequalities* (World Scientific, Singapore, New York, 2011)
2. G. Anastassiou, Self Adjoint Operator Chebyshev-Gr üss type Inequalities (2016)
3. P.L. Čebyšev, Sur les expressions approximatives des intégrales définies par les autres proses entre les mêmes limites. Proc. Math. Soc. Charkov **2**, 93–98 (1882)
4. S.S. Dragomir, Inequalities for functions of selfadjoint operators on Hilbert spaces (2011), ajmaa.org/RGMIA/monographs/InFuncOp.pdf
5. S. Dragomir, *Operator Inequalities of Ostrowski and Trapezoidal Type* (Springer, New York, 2012)
6. A.M. Fink, Bounds on the deviation of a function from its averages. Czech. Math. J. **42**(117), 289–310 (1992)
7. T. Furuta, J. Mićić Hot, J. Pečarić, Y. Seo, Mond-Pečarić Method in Operator Inequalities. Inequalities for bounded self adjoint operators on a Hilbert space, element, Zagreb (2005)
8. G. Grüss, Über das Maximum des absoluten Betrages von $\left[\left(\frac{1}{b-a}\right)\int_a^b f(x)\, g(x)\, dx - \left(\frac{1}{(b-a)^2}\int_a^b f(x)\, dx \int_a^b g(x)\, dx\right)\right]$. Math. Z. **39**, 215–226 (1935)
9. G. Helmberg, *Introduction to Spectral Theory in Hilbert Space* (Wiley, New York, 1969)

Chapter 8
Ultra General Fractional Self Adjoint Operator Representation Formulae and Operator Poincaré and Sobolev and Other Basic Inequalities

We give here many very general fractional self adjoint operator Poincaré and Sobolev type and other basic inner product inequalities to various directions. Initially we give several very general fractional representation formulae in the self adjoint operator sense. Inequalities are based in the self adjoint operator order over a Hilbert space. It follows [5].

8.1 Background

Let A be a selfadjoint linear operator on a complex Hilbert space $(H; \langle \cdot, \cdot \rangle)$. The Gelfand map establishes a $*-$isometrically isomorphism Φ between the set $C(Sp(A))$ of all continuous functions defined on the spectrum of A, denoted $Sp(A)$, and the C^*-algebra $C^*(A)$ generated by A and the identity operator 1_H on H as follows (see e.g. [12, p. 3]):

For any $f, g \in C(Sp(A))$ and any $\alpha, \beta \in \mathbb{C}$ we have

(i) $\Phi(\alpha f + \beta g) = \alpha \Phi(f) + \beta \Phi(g)$;

(ii) $\Phi(fg) = \Phi(f)\Phi(g)$ (the operation composition is on the right) and $\Phi(\overline{f}) = (\Phi(f))^*$;

(iii) $\|\Phi(f)\| = \|f\| := \sup_{t \in Sp(A)} |f(t)|$;

(iv) $\Phi(f_0) = 1_H$ and $\Phi(f_1) = A$, where $f_0(t) = 1$ and $f_1(t) = t$, for $t \in Sp(A)$.

With this notation we define

$$f(A) := \Phi(f), \text{ for all } f \in C(Sp(A)),$$

and we call it the continuous functional calculus for a selfadjoint operator A.

G.A. Anastassiou, *Intelligent Comparisons II: Operator Inequalities and Approximations*, Studies in Computational Intelligence 699,
DOI 10.1007/978-3-319-51475-8_8

If A is a selfadjoint operator and f is a real valued continuous function on $Sp(A)$ then $f(t) \geq 0$ for any $t \in Sp(A)$ implies that $f(A) \geq 0$, i.e. $f(A)$ is a positive operator on H. Moreover, if both f and g are real valued continuous functions on $Sp(A)$ then the following important property holds:

(P) $f(t) \geq g(t)$ for any $t \in Sp(A)$, implies that $f(A) \geq g(A)$ in the operator order of $B(H)$ (the Banach algebra of all bounded linear operators from H into itself).

Equivalently, we use (see [11], pp. 7–8):

Let U be a selfadjoint operator on the complex Hilbert space $(H, \langle \cdot, \cdot \rangle)$ with the spectrum $Sp(U)$ included in the interval $[m, M]$ for some real numbers $m < M$ and $\{E_\lambda\}_\lambda$ be its spectral family.

Then for any continuous function $f : [m, M] \to \mathbb{C}$, it is well known that we have the following spectral representation in terms of the Riemann–Stieljes integral:

$$\langle f(U) x, y \rangle = \int_{m-0}^{M} f(\lambda) d(\langle E_\lambda x, y \rangle), \tag{8.1}$$

for any $x, y \in H$. The function $g_{x,y}(\lambda) := \langle E_\lambda x, y \rangle$ is of bounded variation on the interval $[m, M]$, and

$$g_{x,y}(m-0) = 0 \text{ and } g_{x,y}(M) = \langle x, y \rangle, \tag{8.2}$$

for any $x, y \in H$. Furthermore, it is known that $g_x(\lambda) := \langle E_\lambda x, x \rangle$ is increasing and right continuous on $[m, M]$.

We have also the formula

$$\langle f(U) x, x \rangle = \int_{m-0}^{M} f(\lambda) d(\langle E_\lambda x, x \rangle), \ \forall\, x \in H. \tag{8.3}$$

As a symbol we can write

$$f(U) = \int_{m-0}^{M} f(\lambda) dE_\lambda. \tag{8.4}$$

Above, $m = \min\{\lambda | \lambda \in Sp(U)\} := \min Sp(U)$, $M = \max\{\lambda | \lambda \in Sp(U)\} := \max Sp(U)$. The projections $\{E_\lambda\}_{\lambda \in \mathbb{R}}$, are called the spectral family of A, with the properties:

(a) $E_\lambda \leq E_{\lambda'}$ for $\lambda \leq \lambda'$;

(b) $E_{m-0} = 0_H$ (zero operator), $E_M = 1_H$ (identity operator) and $E_{\lambda+0} = E_\lambda$ for all $\lambda \in \mathbb{R}$.

Furthermore

$$E_\lambda := \varphi_\lambda(U), \forall\, \lambda \in \mathbb{R},$$

is a projection which reduces U, with

$$\varphi_{\lambda}(s) := \begin{cases} 1, \text{ for } -\infty < s \leq \lambda, \\ 0, \text{ for } \lambda < s < +\infty. \end{cases} \tag{8.5}$$

The spectral family $\{E_{\lambda}\}_{\lambda \in \mathbb{R}}$ determines uniquely the self-adjoint operator U and vice versa.

For more on the topic see [13], pp. 256–266, and for more details see there pp. 157–266. See also [10].

Some more basics are given (we follow [11], pp. 1–5):

Let $(H; \langle \cdot, \cdot \rangle)$ be a Hilbert space over $\mathbb{C}$. A bounded linear operator A defined on H is selfjoint, i.e., $A = A^*$, iff $\langle Ax, x \rangle \in \mathbb{R}$, $\forall\, x \in H$, and if A is selfadjoint, then

$$\|A\| = \sup_{x \in H: \|x\|=1} |\langle Ax, x \rangle| . \tag{8.6}$$

Let A, B be selfadjoint operators on H. Then $A \leq B$ iff $\langle Ax, x \rangle \leq \langle Bx, x \rangle, \forall\, x \in H$.

In particular, A is called positive if $A \geq 0$.

Denote by

$$\mathcal{P} := \left\{ \varphi(s) := \sum_{k=0}^{n} \alpha_k s^k | n \geq 0, \alpha_k \in \mathbb{C}, 0 \leq k \leq n \right\}. \tag{8.7}$$

If $A \in \mathcal{B}(H)$ is selfadjoint, and $\varphi(s) \in \mathcal{P}$ has real coefficients, then $\varphi(A)$ is self-adjoint, and

$$\|\varphi(A)\| = \max\{|\varphi(\lambda)|, \lambda \in Sp(A)\}. \tag{8.8}$$

If φ is any function defined on $\mathbb{R}$ we define

$$\|\varphi\|_A := \sup\{|\varphi(\lambda)|, \lambda \in Sp(A)\}. \tag{8.9}$$

If A is selfadjoint operator on Hilbert space H and φ is continuous and given that $\varphi(A)$ is selfadjoint, then $\|\varphi(A)\| = \|\varphi\|_A$. And if φ is a continuous real valued function so it is $|\varphi|$, then $\varphi(A)$ and $|\varphi|(A) = |\varphi(A)|$ are selfadjoint operators (by [11], p. 4, Theorem 7).

Hence it holds

$$\begin{aligned} \||\varphi(A)|\| &= \||\varphi|\|_A = \sup\{||\varphi(\lambda)||, \lambda \in Sp(A)\} \\ &= \sup\{|\varphi(\lambda)|, \lambda \in Sp(A)\} = \|\varphi\|_A = \|\varphi(A)\|, \end{aligned} \tag{8.10}$$

that is

$$\||\varphi(A)|\| = \|\varphi(A)\|. \tag{8.11}$$

For a selfadjoint operator $A \in \mathcal{B}(H)$ which is positive, there exists a unique positive selfadjoint operator $B := \sqrt{A} \in \mathcal{B}(H)$ such that $B^2 = A$, that is $\left(\sqrt{A}\right)^2 = A$. We call B the square root of A.

Let $A \in \mathcal{B}(H)$, then A^*A is selfadjoint and positive. Define the "operator absolute value" $|A| := \sqrt{A^*A}$. If $A = A^*$, then $|A| = \sqrt{A^2}$.

For a continuous real valued function φ we observe the following:

$$|\varphi(A)| \text{ (the functional absolute value) } = \int_{m-0}^{M} |\varphi(\lambda)| \, dE_\lambda =$$

$$\int_{m-0}^{M} \sqrt{(\varphi(\lambda))^2} dE_\lambda = \sqrt{(\varphi(A))^2} = |\varphi(A)| \text{ (operator absolute value)},$$

where A is a selfadjoint operator.

That is we have

$$|\varphi(A)| \text{ (functional absolute value) } = |\varphi(A)| \text{ (operator absolute value)}. \quad (8.12)$$

8.2 Main Results

Let A be a selfadjoint operator in the Hilbert space H with the spectrum $Sp(A) \subseteq [m, M]$, $m < M$; $m, M \in \mathbb{R}$.

In the next we obtain many very general fractional operator representation formulae, and many very general fractional operator Poincaré and Sobolev type inequalities, and many other basic fractional operator inner product inequalities, in the operator order of $\mathcal{B}(H)$ (the Banach algebra of all bounded linear operators from H into itself). All of our functions next in this chapter are real valued.

We mention the following general Taylor formula

Theorem 8.1 ([2], p. 400) *Let $f, f', \ldots, f^{(n)}$; g, g' be continuous from $[m, M]$ into $\mathbb{R}$, $n \in \mathbb{N}$. Assume $\left(g^{-1}\right)^{(k)}$, $k = 0, 1, \ldots, n$, are continuous. Then*

$$f(\lambda) = f(m) + \sum_{k=1}^{n-1} \frac{\left(f \circ g^{-1}\right)^{(k)}(g(m))}{k!} (g(\lambda) - g(m))^k + R_n(m, \lambda), \quad (8.13)$$

where

$$R_n(m, \lambda) = \frac{1}{(n-1)!} \int_m^\lambda (g(\lambda) - g(s))^{n-1} \left(f \circ g^{-1}\right)^{(n)} (g(s)) \, g'(s) \, ds \quad (8.14)$$

$$= \frac{1}{(n-1)!} \int_{g(m)}^{g(\lambda)} (g(\lambda) - t)^{n-1} \left(f \circ g^{-1}\right)^{(n)} (t) \, dt, \ \forall \lambda \in [m, M].$$

We present the operator representation formula

Theorem 8.2 *Let A be a selfadjoint operator in the Hilbert space H with the spectrum $Sp(A) \subseteq [m, M]$ for some real numbers $m < M$, $\{E_\lambda\}_\lambda$ be its spectral family, I be a closed subinterval of $\mathbb{R}$ with $[m, M] \subset \overset{\circ}{I}$ (the interior of I) and $n \in \mathbb{N}$. We consider $f \in C^n([m, M])$, $g \in C^1([m, M])$, such that there exist $\left(g^{-1}\right)^{(k)}$, $k = 0, 1, ..., n$, that are continuous, where $f, g : I \to \mathbb{R}$.*

Then

$$f(A) = \sum_{k=0}^{n-1} \frac{\left(f \circ g^{-1}\right)^{(k)}(g(m))}{k!} \left(g(A) - g(m) 1_H\right)^k + R_n(f, g, m, M), \tag{8.15}$$

where

$$R_n(f, g, m, M) =$$

$$\frac{1}{(n-1)!} \int_{m-0}^{M} \left(\int_m^\lambda (g(\lambda) - g(s))^{n-1} \left(f \circ g^{-1}\right)^{(n)}(g(s))\, g'(s)\, ds \right) dE_\lambda \tag{8.16}$$

$$= \frac{1}{(n-1)!} \int_{m-0}^{M} \left(\int_{g(m)}^{g(\lambda)} (g(\lambda) - t)^{n-1} \left(f \circ g^{-1}\right)^{(n)}(t)\, dt \right) dE_\lambda.$$

Proof We integrate (8.13) and (8.14) against E_λ to get

$$\int_{m-0}^{M} f(\lambda)\, dE_\lambda = \sum_{k=0}^{n-1} \frac{\left(f \circ g^{-1}\right)^{(k)}(g(m))}{k!} \int_{m-0}^{M} (g(\lambda) - g(m))^k\, dE_\lambda$$

$$+ \int_{m-0}^{M} R_n(m, \lambda)\, dE_\lambda. \tag{8.17}$$

By the spectral representation theorem we obtain

$$f(A) = \sum_{k=0}^{n-1} \frac{\left(f \circ g^{-1}\right)^{(k)}(g(m))}{k!} \left(g(A) - g(m) 1_H\right)^k +$$

$$\int_{m-0}^{M} R_n(m, \lambda)\, dE_\lambda, \tag{8.18}$$

proving the claim. ■

Note 8.3 (*to Theorem 8.2*) By [2], p. 401, if $f^{(k)}(m) = 0$, for $k = 0, 1, ..., n-1$, then $\left(f \circ g^{-1}\right)^{(k)}(g(m)) = 0$, all $k = 0, 1, ..., n-1$. In that case it holds

$$f(A) = R_n(f, g, m, M). \tag{8.19}$$

We need

Definition 8.4 *([3])* Let $\alpha > 0$, $\lceil \alpha \rceil = n$, $\lceil \cdot \rceil$ the ceiling of the number. Here $g \in AC([m, M])$ (absolutely continuous functions) and it is strictly increasing. We assume that $\left(f \circ g^{-1}\right)^{(n)} \circ g \in L_\infty([m, M])$. We define the left generalized g-fractional derivative of f of order α as follows:

$$\left(D_{m+;g}^{\alpha} f\right)(x) = \frac{1}{\Gamma(n-\alpha)} \int_m^x (g(x) - g(t))^{n-\alpha-1} g'(t) \left(f \circ g^{-1}\right)^{(n)} (g(t))\, dt, \tag{8.20}$$

$x \geq m$, where Γ is the gamma function.

If $\alpha \notin \mathbb{N}$, by [3], we have that $D_{m+;g}^{\alpha} f \in C([m, M])$.

We set

$$D_{m+;g}^{n} f(x) := \left(\left(f \circ g^{-1}\right)^{(n)} \circ g\right)(x), \tag{8.21}$$

$$\left(D_{m+;g}^{0} f\right)(x) := f(x), \forall\, x \in [m, M].$$

When $g = id$, then

$$D_{m+;g}^{\alpha} f = D_{m+;id}^{\alpha} f = D_{*m}^{\alpha} f, \tag{8.22}$$

the usual left Caputo fractional derivative [1], p. 270, and [9], p. 50.

We need the following g-left fractional generalized Taylor's formula:

Theorem 8.5 ([3]) *Let g be strictly increasing function and $g \in AC([m, M])$. We assume that $\left(f \circ g^{-1}\right) \in AC^n([g(m), g(M)])$, where $\mathbb{N} \ni n = \lceil \alpha \rceil$, $\alpha > 0$ (it means $\left(f \circ g^{-1}\right)^{(n-1)} \in AC([g(m), g(M)])$, and implies that $f \in C([m, M])$). Also we assume that $\left(f \circ g^{-1}\right)^{(n)} \circ g \in L_\infty([m, M])$. Then*

$$f(\lambda) = f(m) + \sum_{k=1}^{n-1} \frac{\left(f \circ g^{-1}\right)^{(k)} (g(m))}{k!} (g(\lambda) - g(m))^k + \tag{8.23}$$

$$\frac{1}{\Gamma(\alpha)} \int_m^{\lambda} (g(\lambda) - g(t))^{\alpha-1} g'(t) \left(D_{m+;g}^{\alpha} f\right)(t)\, dt,$$

$\forall\, \lambda \in [m, M]$.

Calling $R_{\alpha}^{(1)}(m, \lambda)$ the remainder of (8.23), we get that

$$R_{\alpha}^{(1)}(m, \lambda) = \frac{1}{\Gamma(\alpha)} \int_{g(m)}^{g(\lambda)} (g(\lambda) - z)^{\alpha-1} \left(\left(D_{m+;g}^{\alpha} f\right) \circ g^{-1}\right)(z)\, dz, \tag{8.24}$$

$\forall\, \lambda \in [m, M]$.

$R_{\alpha}^{(1)}(m, \lambda)$ is a continuous function in $\lambda \in [m, M]$.

We present the following operator left fractional representation formula

Theorem 8.6 *Let A be a selfadjoint operator in the Hilbert space H with the spectrum $Sp(A) \subseteq [m, M]$ for some real numbers $m < M$, $\{E_\lambda\}_\lambda$ be its spectral family, I be a closed subinterval of $\mathbb{R}$ with $[m, M] \subset \overset{\circ}{I}$ and $n \in \mathbb{N}$, with $n := \lceil \alpha \rceil$, $\alpha > 0$. Let $f, g : I \to \mathbb{R}$. Assume that g is strictly increasing and $g \in AC([m, M])$, and $(f \circ g^{-1}) \in AC^n([g(m), g(M)])$, and $(f \circ g^{-1})^{(n)} \circ g \in L_\infty([m, M])$. Then*

$$f(A) = \sum_{k=0}^{n-1} \frac{(f \circ g^{-1})^{(k)}(g(m))}{k!} (g(A) - g(m) 1_H)^k + R_\alpha^{(1)}(f, g, \alpha, m, M), \tag{8.25}$$

where

$$R_\alpha^{(1)}(f, g, \alpha, m, M) :=$$

$$\frac{1}{\Gamma(\alpha)} \int_{m-0}^{M} \left(\int_m^\lambda (g(\lambda) - g(t))^{\alpha-1} g'(t) \left(D_{m+;g}^\alpha f\right)(t)\, dt \right) dE_\lambda$$

$$= \frac{1}{\Gamma(\alpha)} \int_{m-0}^{M} \left(\int_{g(m)}^{g(\lambda)} (g(\lambda) - z)^{\alpha-1} \left(\left(D_{m+;g}^\alpha f\right) \circ g^{-1}\right)(z)\, dz \right) dE_\lambda. \tag{8.26}$$

Proof We integrate (8.23) against E_λ to get

$$\int_{m-0}^{M} f(\lambda)\, dE_\lambda = \sum_{k=0}^{n-1} \frac{(f \circ g^{-1})^{(k)}(g(m))}{k!} \int_{m-0}^{M} (g(\lambda) - g(m))^k\, dE_\lambda$$

$$+ \int_{m-0}^{M} R_\alpha^{(1)}(m, \lambda)\, dE_\lambda. \tag{8.27}$$

By the spectral representation theorem we obtain

$$f(A) = \sum_{k=0}^{n-1} \frac{(f \circ g^{-1})^{(k)}(g(m))}{k!} (g(A) - g(m) 1_H)^k +$$

$$\int_{m-0}^{M} R_\alpha^{(1)}(m, \lambda)\, dE_\lambda, \tag{8.28}$$

proving the claim. ■

Note 8.7 (*to Theorem 8.6*) If $(f \circ g^{-1})^{(k)}(g(m)) = 0$, for $k = 0, 1, ..., n-1$, then

$$f(A) = R_\alpha^{(1)}(f, g, \alpha, m, M). \tag{8.29}$$

We need

Definition 8.8 *([3])* Let $\alpha > 0$, $\lceil \alpha \rceil = n$. Here $g \in AC([m, M])$ and it is strictly increasing. We assume that $\left(f \circ g^{-1}\right)^{(n)} \circ g \in L_{\infty}([m, M])$. We define the right generalized g-fractional derivative of f of order α as follows:

$$\left(D_{M-;g}^{\alpha} f\right)(x) = \frac{(-1)^n}{\Gamma(n-\alpha)} \int_x^M (g(t) - g(x))^{n-\alpha-1} g'(t) \left(f \circ g^{-1}\right)^{(n)} (g(t))\, dt, \tag{8.30}$$

all $x \in [m, M]$.

If $\alpha \notin \mathbb{N}$, by [3], we get that $\left(D_{M-;g}^{\alpha} f\right) \in C([m, M])$.

We set

$$\left(D_{M-;g}^{n} f\right)(x) := (-1)^n \left(\left(f \circ g^{-1}\right)^{(n)} \circ g\right)(x), \tag{8.31}$$

$$\left(D_{M-;g}^{0} f\right)(x) := f(x), \forall\, x \in [m, M].$$

When $g = id$, then

$$\left(D_{M-;g}^{\alpha} f\right)(x) = \left(D_{M-;id}^{\alpha} f\right)(x) = \left(D_{M-}^{\alpha} f\right)(x), \tag{8.32}$$

the usual right Caputo fractional derivative, [2], pp. 336–337.

We will use the g-right generalized fractional Taylor's formula:

Theorem 8.9 ([3]) *Let g be strictly increasing function and $g \in AC([m, M])$. We assume that $\left(f \circ g^{-1}\right) \in AC^n([g(m), g(M)])$, where $\mathbb{N} \ni n = \lceil \alpha \rceil$, $\alpha > 0$. Also we assume that $\left(f \circ g^{-1}\right)^{(n)} \circ g \in L_{\infty}([m, M])$. Then*

$$f(\lambda) = f(M) + \sum_{k=1}^{n-1} \frac{\left(f \circ g^{-1}\right)^{(k)} (g(M))}{k!} (g(\lambda) - g(M))^k +$$

$$\frac{1}{\Gamma(\alpha)} \int_{\lambda}^{M} (g(t) - g(\lambda))^{\alpha-1} g'(t) \left(D_{M-;g}^{\alpha} f\right)(t)\, dt, \tag{8.33}$$

all $m \leq \lambda \leq M$.

Calling $R_{\alpha}^{(2)}(M, \lambda)$ the remainder of (8.33), we get that

$$R_{\alpha}^{(2)}(M, \lambda) = \frac{1}{\Gamma(\alpha)} \int_{g(\lambda)}^{g(M)} (z - g(\lambda))^{\alpha-1} \left(\left(D_{M-;g}^{\alpha} f\right) \circ g^{-1}\right)(z)\, dz, \tag{8.34}$$

$\forall\, \lambda \in [m, M]$.

$R_{\alpha}^{(2)}(M, \lambda)$ is a continuous function in $\lambda \in [m, M]$.

We present the following operator right fractional representation formula

Theorem 8.10 *Let A be a selfadjoint operator in the Hilbert space H with the spectrum $Sp(A) \subseteq [m, M]$ for some real numbers $m < M$, $\{E_\lambda\}_\lambda$ be its spectral family, I be a closed subinterval of $\mathbb{R}$ with $[m, M] \subset \overset{\circ}{I}$ and $n \in \mathbb{N}$, with $n := \lceil \alpha \rceil$, $\alpha > 0$. Let $f, g : I \to \mathbb{R}$. Assume that g is strictly increasing and $g \in AC([m, M])$, and $(f \circ g^{-1}) \in AC^n([g(m), g(M)])$, and $(f \circ g^{-1})^{(n)} \circ g \in L_\infty([m, M])$. Then*

$$f(A) = \sum_{k=0}^{n-1} \frac{(f \circ g^{-1})^{(k)}(g(M))}{k!} (g(A) - g(M) 1_H)^k + R_\alpha^{(2)}(f, g, \alpha, m, M), \tag{8.35}$$

where

$$R_\alpha^{(2)}(f, g, \alpha, m, M) :=$$

$$\frac{1}{\Gamma(\alpha)} \int_{m-0}^{M} \left(\int_\lambda^M (g(t) - g(\lambda))^{\alpha-1} g'(t) \left(D_{M-;g}^\alpha f\right)(t)\, dt \right) dE_\lambda$$

$$= \frac{1}{\Gamma(\alpha)} \int_{m-0}^{M} \left(\int_{g(\lambda)}^{g(M)} (z - g(\lambda))^{\alpha-1} \left(\left(D_{M-;g}^\alpha f\right) \circ g^{-1}\right)(z)\, dz \right) dE_\lambda. \tag{8.36}$$

Proof We integrate (8.33) against E_λ to get

$$\int_{m-0}^{M} f(\lambda)\, dE_\lambda = \sum_{k=0}^{n-1} \frac{(f \circ g^{-1})^{(k)}(g(M))}{k!} \int_{m-0}^{M} (g(\lambda) - g(M))^k\, dE_\lambda$$

$$+ \int_{m-0}^{M} R_\alpha^{(2)}(M, \lambda)\, dE_\lambda. \tag{8.37}$$

By the spectral representation theorem we obtain

$$f(A) = \sum_{k=0}^{n-1} \frac{(f \circ g^{-1})^{(k)}(g(M))}{k!} (g(A) - g(M) 1_H)^k +$$

$$\int_{m-0}^{M} R_\alpha^{(2)}(M, \lambda)\, dE_\lambda, \tag{8.38}$$

proving the claim. ■

Note 8.11 (*to Theorem 8.10*) If $(f \circ g^{-1})^{(k)}(g(M)) = 0$, for $k = 0, 1, ..., n-1$, then

$$f(A) = R_\alpha^{(2)}(f, g, \alpha, m, M). \tag{8.39}$$

We make

Background 8.12 ([4]) *Let $g : [m, M] \to \mathbb{R}$ be a strictly increasing function. Let $f \in C^n([m, M])$, $n \in \mathbb{N}$. Assume that $g \in C^1([m, M])$, and $g^{-1} \in C^n([g(m), g(M)])$. Call $l := f \circ g^{-1} : [g(m), g(M)] \to \mathbb{R}$. It is clear that $l, l', ..., l^{(n)}$ are continuous from $[g(m), g(M)]$ into $f([m, M]) \subseteq \mathbb{R}$.*

Let $\nu \geq 1$ such that $[\nu] = n$, $n \in \mathbb{N}$ as above, where $[\cdot]$ is the integral part of the number. Clearly when $0 < \nu < 1$, $[\nu] = 0$.

Next we follow [1], pp. 7–9.

Let $h \in C([g(m), g(M)])$, we define the left Riemann–Liouville fractional integral

$$\left(J_{\nu}^{g(m)} h\right)(z) := \frac{1}{\Gamma(\nu)} \int_{g(m)}^{z} (z-t)^{\nu-1} h(t)\, dt, \tag{8.40}$$

for $g(m) \leq z \leq g(M)$.

We set $J_0^{g(m)} h = h$.

Let $\overline{\alpha} := \nu - [\nu]$ $(0 < \overline{\alpha} < 1)$. We define the subspace $C_{g(m)}^{\nu}([g(m), g(M)])$ of $C^{[\nu]}([g(m), g(M)])$ as

$$C_{g(m)}^{\nu}([g(m), g(M)]) :=$$

$$\left\{h \in C^{[\nu]}([g(m), g(M)]) : J_{1-\overline{\alpha}}^{g(m)} h^{([\nu])} \in C^1([g(m), g(M)])\right\}. \tag{8.41}$$

So let $h \in C_{g(m)}^{\nu}([g(m), g(M)])$; we define the left g-generalized fractional derivative of h of order ν, of Canavati type, over $[g(m), g(M)]$ as

$$D_{g(m)}^{\nu} h := \left(J_{1-\overline{\alpha}}^{g(m)} h^{([\nu])}\right)'. \tag{8.42}$$

Clearly, for $h \in C_{g(m)}^{\nu}([g(m), g(M)])$, there exists

$$\left(D_{g(m)}^{\nu} h\right)(z) = \frac{1}{\Gamma(1-\overline{\alpha})} \frac{d}{dz} \int_{g(m)}^{z} (z-t)^{-\overline{\alpha}} h^{([\nu])}(t)\, dt, \tag{8.43}$$

for all $g(m) \leq z \leq g(M)$.

In particular, when $f \circ g^{-1} \in C_{g(m)}^{\nu}([g(m), g(M)])$ we have that

$$\left(D_{g(m)}^{\nu}\left(f \circ g^{-1}\right)\right)(z) = \frac{1}{\Gamma(1-\overline{\alpha})} \frac{d}{dz} \int_{g(m)}^{z} (z-t)^{-\overline{\alpha}} \left(f \circ g^{-1}\right)^{([\nu])}(t)\, dt, \tag{8.44}$$

for all $g(m) \leq z \leq g(M)$.

We have that

$$D_{g(m)}^{n}\left(f \circ g^{-1}\right) = \left(f \circ g^{-1}\right)^{(n)}, \tag{8.45}$$

and

$$D_{g(m)}^{0}\left(f \circ g^{-1}\right) = f \circ g^{-1}. \tag{8.46}$$

We mention the following left generalized g-fractional, of Canavati type, Taylor's formula:

Theorem 8.13 ([4]) *Let* $f \circ g^{-1} \in C^{\nu}_{g(m)}([g(m), g(M)])$.

(i) *if* $\nu \geq 1$, *then*

$$f(\lambda) = \sum_{k=0}^{[\nu]-1} \frac{\left(f \circ g^{-1}\right)^{(k)}(g(m))}{k!}(g(\lambda) - g(m))^k + \tag{8.47}$$

$$\frac{1}{\Gamma(\nu)} \int_{g(m)}^{g(\lambda)} (g(\lambda) - t)^{\nu-1} \left(D^{\nu}_{g(m)}\left(f \circ g^{-1}\right)\right)(t)\, dt,$$

all $\lambda \in [m, M]$,

(ii) *if* $0 < \nu < 1$, *then*

$$f(\lambda) = \frac{1}{\Gamma(\nu)} \int_{g(m)}^{g(\lambda)} (g(\lambda) - t)^{\nu-1} \left(D^{\nu}_{g(m)}\left(f \circ g^{-1}\right)\right)(t)\, dt, \tag{8.48}$$

all $\lambda \in [m, M]$.

By the change of variable method, see [14], we may rewrite the remainder of (8.47) and (8.48), as

$$R^{(3)}_{\nu}(m, \lambda) := \frac{1}{\Gamma(\nu)} \int_{g(m)}^{g(\lambda)} (g(\lambda) - t)^{\nu-1} \left(D^{\nu}_{g(m)}\left(f \circ g^{-1}\right)\right)(t)\, dt = \tag{8.49}$$

$$\frac{1}{\Gamma(\nu)} \int_{m}^{\lambda} (g(\lambda) - g(s))^{\nu-1} \left(D^{\nu}_{g(m)}\left(f \circ g^{-1}\right)\right)(g(s))\, g'(s)\, ds,$$

all $\lambda \in [m, M]$.

We present the following operator left fractional representation formula.

Theorem 8.14 *Let* A *be a selfadjoint operator in the Hilbert space* H *with the spectrum* $Sp(A) \subseteq [m, M]$ *for some real numbers* $m < M$, $\{E_\lambda\}_\lambda$ *be its spectral family,* I *be a closed subinterval of* $\mathbb{R}$ *with* $[m, M] \subset \overset{\circ}{I}$ *and* $n \in \mathbb{N}$, *with* $n := [\nu]$, $\nu > 0$. *Let* $f, g : I \to \mathbb{R}$. *Assume that* $g : [m, M] \to \mathbb{R}$ *is strictly increasing function,* $f \in C^n([m, M])$, $g \in C^1([m, M])$, *and* $g^{-1} \in C^n([g(m), g(M)])$. *Suppose also that* $f \circ g^{-1} \in C^{\nu}_{g(m)}([g(m), g(M)])$. *Then*

(i) *if* $\nu \geq 1$, *then*

$$f(A) = \sum_{k=0}^{[\nu]-1} \frac{\left(f \circ g^{-1}\right)^{(k)}(g(m))}{k!}(g(A) - g(m)1_H)^k + R^{(3)}_{\nu}(f, g, \nu, m, M), \tag{8.50}$$

(ii) if $0 < \nu < 1$, *then*

$$f(A) = R_{\nu}^{(3)}(f, g, \nu, m, M). \tag{8.51}$$

Here it is

$$R_{\nu}^{(3)}(f, g, \nu, m, M) :=$$

$$\frac{1}{\Gamma(\nu)} \int_{m-0}^{M} \left(\int_{g(m)}^{g(\lambda)} (g(\lambda) - t)^{\nu-1} \left(D_{g(m)}^{\nu} \left(f \circ g^{-1} \right) \right)(t)\, dt \right) dE_{\lambda} =$$

$$\frac{1}{\Gamma(\nu)} \int_{m-0}^{M} \left(\int_{m}^{\lambda} (g(\lambda) - g(s))^{\nu-1} \left(D_{g(m)}^{\nu} \left(f \circ g^{-1} \right) \right)(g(s))\, g'(s)\, ds \right) dE_{\lambda}. \tag{8.52}$$

Proof We integrate (8.47) and (8.48) against E_{λ} and use the spectral representation theorem, as in Theorem 8.6. ■

Note 8.15 If $\nu \geq 1$ and $f^{(k)}(m) = 0$, then $\left(f \circ g^{-1} \right)^{(k)}(g(m)) = 0$, all $k = 0, 1, ..., [\nu] - 1$, (see [2], p. 401), and

$$f(A) = R_{\nu}^{(3)}(f, g, \nu, m, M). \tag{8.53}$$

We need

Background 8.16 *Let* g, f, l, ν, n, h *as in Background 8.12. Here we follow [2], pp. 345–348.*

We define the right Riemann-Liouville fractional integral as

$$\left(J_{g(M)-}^{\nu} h \right)(z) := \frac{1}{\Gamma(\nu)} \int_{z}^{g(M)} (t - z)^{\nu-1} h(t)\, dt, \tag{8.54}$$

for $g(m) \leq z \leq g(M)$.

We set $J_{g(M)-}^{0} h = h$.

Let $\overline{\alpha} := \nu - [\nu]$ $(0 < \overline{\alpha} < 1)$. *We define the subspace* $C_{g(M)-}^{\nu}([g(m), g(M)])$ *of* $C^{[\nu]}([g(m), g(M)])$ *as*

$$C_{g(M)-}^{\nu}([g(m), g(M)]) :=$$

$$\left\{ h \in C^{[\nu]}([g(m), g(M)]) : J_{g(M)-}^{1-\overline{\alpha}} h^{([\nu])} \in C^{1}([g(m), g(M)]) \right\}. \tag{8.55}$$

So let $h \in C_{g(M)-}^{\nu}([g(m), g(M)])$; *we define the right* g*-generalized fractional derivative of* h *of order* ν, *of Canavati type, over* $[g(m), g(M)]$ *as*

$$D_{g(M)-}^{\nu} h := (-1)^{n-1} \left(J_{g(M)-}^{1-\overline{\alpha}} h^{([\nu])} \right)'. \tag{8.56}$$

Clearly, for $h \in C^{\nu}_{g(M)-}([g(m), g(M)])$, *there exists*

$$\left(D^{\nu}_{g(M)-}h\right)(z) = \frac{(-1)^{n-1}}{\Gamma(1-\overline{\alpha})} \frac{d}{dz} \int_{z}^{g(M)} (t-z)^{-\overline{\alpha}} h^{([\nu])}(t)\, dt, \tag{8.57}$$

for all $g(m) \leq z \leq g(M)$.

In particular, when $f \circ g^{-1} \in C^{\nu}_{g(M)-}([g(m), g(M)])$ *we have that*

$$\left(D^{\nu}_{g(M)-}\left(f \circ g^{-1}\right)\right)(z) = \frac{(-1)^{n-1}}{\Gamma(1-\overline{\alpha})} \frac{d}{dz} \int_{z}^{g(M)} (t-z)^{-\overline{\alpha}} \left(f \circ g^{-1}\right)^{([\nu])}(t)\, dt, \tag{8.58}$$

for all $g(m) \leq z \leq g(M)$.

We get that

$$\left(D^{n}_{g(M)-}\left(f \circ g^{-1}\right)\right)(z) = (-1)^{n}\left(f \circ g^{-1}\right)^{(n)}(z), \tag{8.59}$$

and

$$\left(D^{0}_{g(M)-}\left(f \circ g^{-1}\right)\right)(z) = \left(f \circ g^{-1}\right)(z), \tag{8.60}$$

all $z \in [g(m), g(M)]$.

We need the following right generalized g-fractional, of Canavati type, Taylor's formula:

Theorem 8.17 ([4]) *Let* $f \circ g^{-1} \in C^{\nu}_{g(M)-}([g(m), g(M)])$.

(i) *if* $\nu \geq 1$, *then*

$$f(\lambda) = \sum_{k=0}^{[\nu]-1} \frac{\left(f \circ g^{-1}\right)^{(k)}(g(M))}{k!}(g(\lambda) - g(M))^{k} +$$

$$\frac{1}{\Gamma(\nu)} \int_{g(\lambda)}^{g(M)} (t - g(\lambda))^{\nu-1}\left(D^{\nu}_{g(M)-}\left(f \circ g^{-1}\right)\right)(t)\, dt, \tag{8.61}$$

all $m \leq \lambda \leq M$,

(ii) *if* $0 < \nu < 1$, *we get*

$$f(\lambda) = \frac{1}{\Gamma(\nu)} \int_{g(\lambda)}^{g(M)} (t - g(\lambda))^{\nu-1}\left(D^{\nu}_{g(M)-}\left(f \circ g^{-1}\right)\right)(t)\, dt, \tag{8.62}$$

all $m \leq \lambda \leq M$.

By change of variable, see [14], we may rewrite the remainder of (8.61) and (8.62), as

$$R_{\nu}^{(4)}(M,\lambda) := \frac{1}{\Gamma(\nu)} \int_{g(\lambda)}^{g(M)} (t - g(\lambda))^{\nu-1} \left(D_{g(M)-}^{\nu} \left(f \circ g^{-1} \right) \right)(t)\, dt =$$

$$\frac{1}{\Gamma(\nu)} \int_{\lambda}^{M} (g(s) - g(\lambda))^{\nu-1} \left(D_{g(M)-}^{\nu} \left(f \circ g^{-1} \right) \right)(g(s))\, g'(s)\, ds, \tag{8.63}$$

all $m \leq \lambda \leq M$.

We present the following operator right fractional representation formula

Theorem 8.18 *Let A be a selfadjoint operator in the Hilbert space H with the spectrum $Sp(A) \subseteq [m, M]$ for some real numbers $m < M$, $\{E_\lambda\}_\lambda$ be its spectral family, I be a closed subinterval of $\mathbb{R}$ with $[m, M] \subset \overset{\circ}{I}$ and $n \in \mathbb{N}$, with $n := \lceil \nu \rceil$, $\nu > 0$. Let $f, g : I \to \mathbb{R}$. Assume that $g : [m, M] \to \mathbb{R}$ is strictly increasing function, $f \in C^n([m, M])$, $g \in C^1([m, M])$, and $g^{-1} \in C^n([g(m), g(M)])$. Suppose also that $f \circ g^{-1} \in C_{g(M)-}^{\nu}([g(m), g(M)])$. Then*

(i) if $\nu \geq 1$, then

$$f(A) = \sum_{k=0}^{\lceil \nu \rceil - 1} \frac{\left(f \circ g^{-1} \right)^{(k)}(g(M))}{k!} (g(A) - g(M) 1_H)^k + R_{\nu}^{(4)}(f, g, \nu, m, M), \tag{8.64}$$

(ii) if $0 < \nu < 1$, then

$$f(A) = R_{\nu}^{(4)}(f, g, \nu, m, M). \tag{8.65}$$

Here it is

$$R_{\nu}^{(4)}(f, g, \nu, m, M) :=$$

$$\frac{1}{\Gamma(\nu)} \int_{m-0}^{M} \left(\int_{g(\lambda)}^{g(M)} (t - g(\lambda))^{\nu-1} \left(D_{g(M)-}^{\nu} \left(f \circ g^{-1} \right) \right)(t)\, dt \right) dE_\lambda =$$

$$\frac{1}{\Gamma(\nu)} \int_{m-0}^{M} \left(\int_{\lambda}^{M} (g(s) - g(\lambda))^{\nu-1} \left(D_{g(M)-}^{\nu} \left(f \circ g^{-1} \right) \right)(g(s))\, g'(s)\, ds \right) dE_\lambda. \tag{8.66}$$

Proof We integrate (8.61) and (8.62) against E_λ and use the spectral representation theorem, as in Theorem 8.10. ■

Note 8.19 If $\nu \geq 1$ and $f^{(k)}(M) = 0$, then $\left(f \circ g^{-1} \right)^{(k)}(g(M)) = 0$, all $k = 0, 1, \ldots, \lceil \nu \rceil - 1$, (see [2], p. 401), and

$$f(A) = R_{\nu}^{(4)}(f, g, \nu, m, M). \tag{8.67}$$

We need

Background 8.20 *Let* $f : [m, M] \to \mathbb{R} : f^{(\overline{m})} \in L_\infty([m, M])$, *the left Caputo fractional derivative ([9], p. 50) of order* $\alpha \notin \mathbb{N}$, $\alpha > 0$, $\overline{m} = \lceil \alpha \rceil$ ($\lceil \cdot \rceil$ *ceiling) is defined as follows:*

$$\left(D_{*m}^{\alpha} f\right)(x) = \frac{1}{\Gamma(\overline{m} - \alpha)} \int_m^x (x - t)^{\overline{m} - \alpha - 1} f^{(\overline{m})}(t)\, dt, \tag{8.68}$$

$\forall\, x \in [m, M]$.

Let $n \in \mathbb{N}$, *we denote*

$$D_{*m}^{n\alpha} = D_{*m}^{\alpha} D_{*m}^{\alpha} \ldots D_{*m}^{\alpha} \text{ (}n\text{-times)}. \tag{8.69}$$

Let us assume now that

$$D_{*m}^{k\alpha} f \in C([m, M]),\ k = 0, 1, \ldots, n + 1;\ n \in \mathbb{N},\ 0 < \alpha \leq 1. \tag{8.70}$$

By [6, 15], we mention the following generalized fractional Caputo type Taylor's formula:

$$f(\lambda) = \sum_{i=0}^{n} \frac{(\lambda - m)^{i\alpha}}{\Gamma(i\alpha + 1)} \left(D_{*m}^{i\alpha} f\right)(m) +$$

$$\frac{1}{\Gamma((n + 1)\alpha)} \int_m^{\lambda} (\lambda - t)^{(n+1)\alpha - 1} \left(D_{*m}^{(n+1)\alpha} f\right)(t)\, dt, \tag{8.71}$$

$\forall\, \lambda \in [m, M]$.

We give the following operator left fractional representation formula

Theorem 8.21 *Let* A *be a selfadjoint operator in the Hilbert space* H *with the spectrum* $Sp(A) \subseteq [m, M]$ *for some real numbers* $m < M$, $\{E_\lambda\}_\lambda$ *be its spectral family,* I *be a closed subinterval of* $\mathbb{R}$ *with* $[m, M] \subset \overset{\circ}{I}$. *Here* $f : I \to \mathbb{R}$. *Furthermore assume that* $f' \in L_\infty([m, M])$, *and* $D_{*m}^{k\alpha} f \in C([m, M])$, $k = 0, 1, \ldots, n + 1$; $n \in \mathbb{N}$, $0 < \alpha \leq 1$. *Then*

$$f(A) = \sum_{i=0}^{n} \frac{\left(D_{*m}^{i\alpha} f\right)(m)}{\Gamma(i\alpha + 1)} (A - m1_H)^{i\alpha} +$$

$$\frac{1}{\Gamma((n + 1)\alpha)} \int_{m-0}^{M} \left(\int_m^{\lambda} (\lambda - t)^{(n+1)\alpha - 1} \left(D_{*m}^{(n+1)\alpha} f\right)(t)\, dt \right) dE_\lambda. \tag{8.72}$$

Proof We use (8.71) and the spectral representation theorem, as in Theorem 8.6. ■

Note 8.22 *(to Theorem 8.21)* If $\left(D_{*m}^{i\alpha}f\right)(m)=0$, $i=0,1,...,n$, then

$$f(A)=\frac{1}{\Gamma((n+1)\alpha)}\int_{m-0}^{M}\left(\int_{m}^{\lambda}(\lambda-t)^{(n+1)\alpha-1}\left(D_{*m}^{(n+1)\alpha}f\right)(t)\,dt\right)dE_{\lambda}. \tag{8.73}$$

We need

Background 8.23 *The right Caputo fractional derivative of order $\alpha>0$, $\overline{m}=\lceil\alpha\rceil$, $f\in AC^{\overline{m}}([m,M])$ is defined as follows (see [2], p. 336):*

$$\left(D_{M-}^{\alpha}f\right)(x)=\frac{(-1)^{\overline{m}}}{\Gamma(\overline{m}-\alpha)}\int_{x}^{M}(z-x)^{\overline{m}-\alpha-1}f^{(\overline{m})}(z)\,dz, \tag{8.74}$$

$\forall\, x\in[m,M]$, *with*

$$D_{M-}^{\overline{m}}f(x):=(-1)^{\overline{m}}f^{(\overline{m})}(x). \tag{8.75}$$

Denote by

$$D_{M-}^{n\alpha}=D_{M-}^{\alpha}D_{M-}^{\alpha}...D_{M-}^{\alpha}\text{ (n-times)},\ n\in\mathbb{N}. \tag{8.76}$$

We need the following right generalized fractional Taylor's formula

Theorem 8.24 ([3]) *Suppose that $f\in AC([m,M])$ and $D_{M-}^{k\alpha}f\in C([m,M])$, for $k=0,1,...,n+1$, where $0<\alpha\leq1$. Then*

$$f(\lambda)=\sum_{i=0}^{n}\frac{(M-\lambda)^{i\alpha}}{\Gamma(i\alpha+1)}\left(D_{M-}^{i\alpha}f\right)(M)+$$

$$\frac{1}{\Gamma((n+1)\alpha)}\int_{\lambda}^{M}(z-\lambda)^{(n+1)\alpha-1}\left(D_{M-}^{(n+1)\alpha}f\right)(z)\,dz, \tag{8.77}$$

$\forall\,\lambda\in[m,M]$.

We give the following operator right fractional representation formula.

Theorem 8.25 *Let A be a selfadjoint operator in the Hilbert space H with the spectrum $Sp(A)\subseteq[m,M]$ for some real numbers $m<M$, $\{E_\lambda\}_\lambda$ be its spectral family, I be a closed subinterval of $\mathbb{R}$ with $[m,M]\subset\overset{\circ}{I}$. Here $f:I\to\mathbb{R}$. Furthermore assume that $f\in AC([m,M])$, and $D_{M-}^{k\alpha}f\in C([m,M])$, for $k=0,1,...,n+1$, where $0<\alpha\leq1$. Then*

$$f(A)=\sum_{i=0}^{n}\frac{(M1_H-A)^{i\alpha}}{\Gamma(i\alpha+1)}\left(D_{M-}^{i\alpha}f\right)(M)+$$

$$\frac{1}{\Gamma((n+1)\alpha)}\int_{m-0}^{M}\left(\int_{\lambda}^{M}(z-\lambda)^{(n+1)\alpha-1}\left(D_{M-}^{(n+1)\alpha}f\right)(z)\,dz\right)dE_{\lambda}. \tag{8.78}$$

Proof Use of (8.77) and spectral representation theorem, as in Theorem 8.21. ■

Note 8.26 (*to Theorem 8.25*) If $\left(D_{M-}^{i\alpha} f\right)(M) = 0$, $i = 0, 1, \ldots, n$, then

$$f(A) = \frac{1}{\Gamma((n+1)\alpha)} \int_{m-0}^{M} \left(\int_{\lambda}^{M} (z-\lambda)^{(n+1)\alpha-1} \left(D_{M-}^{(n+1)\alpha} f\right)(z)\, dz \right) dE_{\lambda}. \tag{8.79}$$

Background 8.27 ([3]) *Denote by* $(\alpha > 0)$

$$D_{m+;g}^{n\alpha} := D_{m+;g}^{\alpha} D_{m+;g}^{\alpha} \ldots D_{m+;g}^{\alpha} \text{ (n-times)}, \; n \in \mathbb{N}. \tag{8.80}$$

By convention $D_{m+;g}^{0} = I$ *(identity operator).*

We need the following left general fractional Taylor's formula.

Theorem 8.28 ([3]) *Let g be strictly increasing and* $g \in AC([m, M])$. *Suppose that* $F_k := D_{m+;g}^{k\alpha} f$, *for* $k = 0, 1, \ldots, n+1$, *fulfill:* $F_k \circ g^{-1} \in AC([g(m), g(M)])$ *and* $\left(F_k \circ g^{-1}\right)' \circ g \in L_{\infty}([m, M])$, *where* $0 < \alpha \leq 1$. *Then*

$$f(\lambda) = \sum_{i=0}^{n} \frac{(g(\lambda) - g(m))^{i\alpha}}{\Gamma(i\alpha+1)} \left(D_{m+;g}^{i\alpha} f\right)(m) +$$

$$\frac{1}{\Gamma((n+1)\alpha)} \int_{m}^{\lambda} (g(\lambda) - g(t))^{(n+1)\alpha-1} g'(t) \left(D_{m+;g}^{(n+1)\alpha} f\right)(t)\, dt, \tag{8.81}$$

$\forall \lambda \in [m, M]$.

We give the following operator general left fractional representation formula.

Theorem 8.29 *Let A be a selfadjoint operator in the Hilbert space H with the spectrum* $Sp(A) \subseteq [m, M]$ *for some real numbers* $m < M$, $\{E_\lambda\}_\lambda$ *be its spectral family, I be a closed subinterval of* $\mathbb{R}$ *with* $[m, M] \subset \overset{\circ}{I}$. *Here* $f, g : I \to \mathbb{R}$. *Furthermore we assume that g is strictly increasing and* $g \in AC([m, M])$. *Suppose that* $F_k := D_{m+;g}^{k\alpha} f$, *for* $k = 0, 1, \ldots, n+1$, *fulfill:* $F_k \circ g^{-1} \in AC([g(m), g(M)])$, *and* $\left(F_k \circ g^{-1}\right)' \circ g \in L_{\infty}([m, M])$, *where* $0 < \alpha \leq 1$. *Then*

$$f(A) = \sum_{i=0}^{n} \frac{\left(D_{m+;g}^{i\alpha} f\right)(m)}{\Gamma(i\alpha+1)} (g(A) - g(m) 1_H)^{i\alpha} +$$

$$\frac{1}{\Gamma((n+1)\alpha)} \int_{m-0}^{M} \left(\int_{m}^{\lambda} (g(\lambda) - g(t))^{(n+1)\alpha-1} g'(t) \left(D_{m+;g}^{(n+1)\alpha} f\right)(t)\, dt \right) dE_{\lambda}. \tag{8.82}$$

Proof Use of (8.81) and spectral representation theorem. ■

Note 8.30 (*to Theorem 8.29*) If $\left(D^{i\alpha}_{m+;g}f\right)(m) = 0$, $i = 0, 1, ..., n$, then

$$f(A) = \frac{1}{\Gamma((n+1)\alpha)} \cdot$$

$$\int_{m-0}^{M} \left(\int_{m}^{\lambda} (g(\lambda) - g(t))^{(n+1)\alpha - 1} g'(t) \left(D^{(n+1)\alpha}_{m+;g}f\right)(t)\, dt \right) dE_{\lambda}. \tag{8.83}$$

We need

Background 8.31 ([3]) *Denote by* $(\alpha > 0)$

$$D^{n\alpha}_{M-;g} := D^{\alpha}_{M-;g} D^{\alpha}_{M-;g} ... D^{\alpha}_{M-;g} \text{ (n-times)},\ n \in \mathbb{N}. \tag{8.84}$$

By convention $D^{0}_{M-;g} = I$ *(identity operator).*

We need the following right general fractional Taylor's formula

Theorem 8.32 ([3]) *Let* g *be strictly increasing and* $g \in AC([m, M])$. *Suppose that* $F_k := D^{k\alpha}_{M-;g}f$, *for* $k = 0, 1, ..., n+1$, *fulfill:* $F_k \circ g^{-1} \in AC([g(m), g(M)])$ *and* $\left(F_k \circ g^{-1}\right)' \circ g \in L_{\infty}([m, M])$, *where* $0 < \alpha \leq 1$. *Then*

$$f(\lambda) = \sum_{i=0}^{n} \frac{(g(M) - g(\lambda))^{i\alpha}}{\Gamma(i\alpha + 1)} \left(D^{i\alpha}_{M-;g}f\right)(M) +$$

$$\frac{1}{\Gamma((n+1)\alpha)} \int_{\lambda}^{M} (g(t) - g(\lambda))^{(n+1)\alpha - 1} g'(t) \left(D^{(n+1)\alpha}_{M-;g}f\right)(t)\, dt, \tag{8.85}$$

$\forall\, \lambda \in [m, M]$.

We give the following operator general right fractional representation formula

Theorem 8.33 *Let* A *be a selfadjoint operator in the Hilbert space* H *with the spectrum* $Sp(A) \subseteq [m, M]$ *for some real numbers* $m < M$, $\{E_{\lambda}\}_{\lambda}$ *be its spectral family,* I *be a closed subinterval of* $\mathbb{R}$ *with* $[m, M] \subset \overset{\circ}{I}$. *Here* $f, g : I \to \mathbb{R}$. *Furthermore we assume that* g *is strictly increasing and* $g \in AC([m, M])$. *Suppose that* $F_k := D^{k\alpha}_{M-;g}f$, *for* $k = 0, 1, ..., n+1$, *fulfill:* $F_k \circ g^{-1} \in AC([g(m), g(M)])$, *and* $\left(F_k \circ g^{-1}\right)' \circ g \in L_{\infty}([m, M])$, *where* $0 < \alpha \leq 1$. *Then*

$$f(A) = \sum_{i=0}^{n} \frac{\left(D^{i\alpha}_{M-;g}f\right)(M)}{\Gamma(i\alpha + 1)} (g(M) 1_H - g(A))^{i\alpha} +$$

$$\frac{1}{\Gamma((n+1)\alpha)} \int_{m-0}^{M} \left(\int_{\lambda}^{M} (g(t) - g(\lambda))^{(n+1)\alpha - 1} g'(t) \left(D^{(n+1)\alpha}_{M-;g}f\right)(t)\, dt \right) dE_{\lambda}. \tag{8.86}$$

Proof Use of (8.85) and spectral representation theorem. ■

Note 8.34 (*to Theorem 8.33*) If $\left(D_{M-;g}^{i\alpha} f\right)(M) = 0$, $i = 0, 1, ..., n$, then

$$f(A) = \frac{1}{\Gamma((n+1)\alpha)} \cdot$$

$$\int_{m-0}^{M} \left(\int_{\lambda}^{M} (g(t) - g(\lambda))^{(n+1)\alpha - 1} g'(t) \left(D_{M-;g}^{(n+1)\alpha} f\right)(t)\, dt \right) dE_{\lambda}. \tag{8.87}$$

We need

Background 8.35 ([4]) *Denote by*

$$D_{g(m)}^{\overline{m}\nu} = D_{g(m)}^{\nu} D_{g(m)}^{\nu} ... D_{g(m)}^{\nu} \; (\overline{m}\text{-times}),\; \overline{m} \in \mathbb{N}. \tag{8.88}$$

We will use the left fractional Taylor's formula

Theorem 8.36 ([4]) *Let* $0 < \nu < 1$.
Assume that $\left(D_{g(m)}^{i\nu} \left(f \circ g^{-1}\right)\right) \in C_{g(m)}^{\nu}([g(m), g(M)])$, $i = 0, 1, ..., \overline{m}$. *Assume also that* $\left(D_{g(m)}^{(\overline{m}+1)\nu} \left(f \circ g^{-1}\right)\right) \in C([g(m), g(M)])$. *Then*

$$f(\lambda) = \frac{1}{\Gamma((\overline{m}+1)\nu)} \int_{g(m)}^{g(\lambda)} (g(\lambda) - z)^{(\overline{m}+1)\nu - 1} \left(D_{g(m)}^{(\overline{m}+1)\nu} \left(f \circ g^{-1}\right)\right)(z)\, dz$$

$$= \frac{1}{\Gamma((\overline{m}+1)\nu)} \int_{m}^{\lambda} (g(\lambda) - g(s))^{(\overline{m}+1)\nu - 1} \left(D_{g(m)}^{(\overline{m}+1)\nu} \left(f \circ g^{-1}\right)\right)(g(s))\, g'(s)\, ds, \tag{8.89}$$

all $m \leq \lambda \leq M$.

We present the operator left fractional representation formula

Theorem 8.37 *Let* A *be a selfadjoint operator in the Hilbert space* H *with the spectrum* $Sp(A) \subseteq [m, M]$ *for some real numbers* $m < M$, $\{E_{\lambda}\}_{\lambda}$ *be its spectral family,* I *be a closed subinterval on* $\mathbb{R}$ *with* $[m, M] \subset \overset{\circ}{I}$, *and* $0 < \nu < 1$. *Let* $f, g : I \to \mathbb{R}$. *Assume that* $g : [m, M] \to \mathbb{R}$ *is strictly increasing function,* $f \in C^1([m, M])$, $g \in C^1([m, M])$, *and* $g^{-1} \in C^1([g(m), g(M)])$. *Furthermore we suppose that* $\left(D_{g(m)}^{i\nu} \left(f \circ g^{-1}\right)\right) \in C_{g(m)}^{\nu}([g(m), g(M)])$, $i = 0, 1, ..., \overline{m}$, *and* $\left(D_{g(m)}^{(\overline{m}+1)\nu} \left(f \circ g^{-1}\right)\right) \in C([g(m), g(M)])$. *Then*

$$f(A) = \frac{1}{\Gamma((\overline{m}+1)\nu)} \cdot$$

$$\int_{m-0}^{M}\left(\int_{g(m)}^{g(\lambda)}(g(\lambda)-z)^{(\overline{m}+1)\nu-1}\left(D_{g(m)}^{(\overline{m}+1)\nu}\left(f\circ g^{-1}\right)\right)(z)\,dz\right)dE_{\lambda} \tag{8.90}$$

$$=\frac{1}{\Gamma((\overline{m}+1)\nu)}\cdot$$

$$\int_{m-0}^{M}\left(\int_{m}^{\lambda}(g(\lambda)-g(s))^{(\overline{m}+1)\nu-1}\left(D_{g(m)}^{(\overline{m}+1)\nu}\left(f\circ g^{-1}\right)\right)((s))\,g'(s)\,ds\right)dE_{\lambda}.$$

Proof Use of (8.89). ∎

We need

Background 8.38 ([4]) *Denote by*

$$D_{g(M)-}^{\overline{m}\nu}=D_{g(M)-}^{\nu}D_{g(M)-}^{\nu}...D_{g(M)-}^{\nu}\ (\overline{m}\text{-times}),\ \overline{m}\in\mathbb{N}. \tag{8.91}$$

We will use the right fractional Taylor's formula

Theorem 8.39 ([4]) *Let* $0<\nu<1$.
Assume that $\left(D_{g(M)-}^{i\nu}\left(f\circ g^{-1}\right)\right)\in C_{g(M)-}^{\nu}([g(m),g(M)])$, *for all* $i=0,1,\ldots,\overline{m}$. *Assume also that* $\left(D_{g(M)-}^{(\overline{m}+1)\nu}\left(f\circ g^{-1}\right)\right)\in C([g(m),g(M)])$. *Then*

$$f(\lambda)=\frac{1}{\Gamma((\overline{m}+1)\nu)}\int_{g(\lambda)}^{g(M)}(z-g(\lambda))^{(\overline{m}+1)\nu-1}\left(D_{g(M)-}^{(\overline{m}+1)\nu}\left(f\circ g^{-1}\right)\right)(z)\,dz$$

$$=\frac{1}{\Gamma((\overline{m}+1)\nu)}\int_{\lambda}^{M}(g(s)-g(\lambda))^{(\overline{m}+1)\nu-1}\left(D_{g(M)-}^{(\overline{m}+1)\nu}\left(f\circ g^{-1}\right)\right)(g(s))\,g'(s)\,ds, \tag{8.92}$$

all $m\leq\lambda\leq M$.

We present the operator right fractional representation formula

Theorem 8.40 *Let A be a selfadjoint operator in the Hilbert space H with the spectrum* $Sp(A)\subseteq[m,M]$ *for some real numbers* $m<M$, $\{E_{\lambda}\}_{\lambda}$ *be its spectral family, I be a closed subinterval on* $\mathbb{R}$ *with* $[m,M]\subset\overset{\circ}{I}$, *and* $0<\nu<1$. *Let* $f,g:I\to\mathbb{R}$. *Assume that* $g:[m,M]\to\mathbb{R}$ *is strictly increasing function,* $f\in C^{1}([m,M])$, $g\in C^{1}([m,M])$, *and* $g^{-1}\in C^{1}([g(m),g(M)])$. *Furthermore we suppose that* $\left(D_{g(M)-}^{i\nu}\left(f\circ g^{-1}\right)\right)\in C_{g(M)-}^{\nu}([g(m),g(M)])$, *for all* $i=0,1,\ldots,\overline{m}$, *and* $\left(D_{g(M)-}^{(\overline{m}+1)\nu}\left(f\circ g^{-1}\right)\right)\in C([g(m),g(M)])$. *Then*

$$f(A)=\frac{1}{\Gamma((\overline{m}+1)\nu)}\cdot$$

$$\int_{m-0}^{M}\left(\int_{g(\lambda)}^{g(M)}(z-g(\lambda))^{(\overline{m}+1)\nu-1}\left(D_{g(M)-}^{(\overline{m}+1)\nu}\left(f\circ g^{-1}\right)\right)(z)\,dz\right)dE_{\lambda}$$

$$=\frac{1}{\Gamma\left((\overline{m}+1)\nu\right)}\cdot$$

$$\int_{m-0}^{M}\left(\int_{\lambda}^{M}(g(s)-g(\lambda))^{(\overline{m}+1)\nu-1}\left(D_{g(M)-}^{(\overline{m}+1)\nu}\left(f\circ g^{-1}\right)\right)(g(s))\,g'(s)\,ds\right)dE_{\lambda}. \tag{8.93}$$

Proof Use of (8.92). ∎

Note 8.41 From now on in this chapter let $p,q>1:\frac{1}{p}+\frac{1}{q}=1$.

We make

Remark 8.42 (to Theorems 8.1 and 8.2) Assume $f^{(k)}(m)=0$, for $k=0,1,...,n-1$, then $\left(f\circ g^{-1}\right)^{(k)}(g(m))=0$, all $k=0,1,...,n-1$, and

$$f(\lambda)=\frac{1}{(n-1)!}\int_{g(m)}^{g(\lambda)}(g(\lambda)-t)^{n-1}\left(f\circ g^{-1}\right)^{(n)}(t)\,dt, \tag{8.94}$$

$\forall\,\lambda\in[m,M]$. Hence, if $g(\lambda)\geq g(m)$, we have

$$|f(\lambda)|\leq\frac{1}{(n-1)!}\int_{g(m)}^{g(\lambda)}(g(\lambda)-t)^{n-1}\left|\left(f\circ g^{-1}\right)^{(n)}(t)\right|dt$$

$$\leq\frac{1}{(n-1)!}\left(\int_{g(m)}^{g(\lambda)}(g(\lambda)-t)^{p(n-1)}\,dt\right)^{\frac{1}{p}}\left(\int_{g(m)}^{g(\lambda)}\left|\left(f\circ g^{-1}\right)^{(n)}(t)\right|^{q}dt\right)^{\frac{1}{q}}$$

$$=\frac{1}{(n-1)!}\frac{(g(\lambda)-g(m))^{(n-1)+\frac{1}{p}}}{(p(n-1)+1)^{\frac{1}{p}}}\left(\int_{g(m)}^{g(\lambda)}\left|\left(f\circ g^{-1}\right)^{(n)}(t)\right|^{q}dt\right)^{\frac{1}{q}}=$$

$$\frac{1}{(n-1)!}\frac{(g(\lambda)-g(m))^{n-\frac{1}{q}}}{(p(n-1)+1)^{\frac{1}{p}}}\left\|\left(f\circ g^{-1}\right)^{(n)}\right\|_{q,[g(m),g(\lambda)]}. \tag{8.95}$$

We have proved that (if $g(\lambda)\geq g(m)$)

$$\left|\int_{g(m)}^{g(\lambda)}(g(\lambda)-t)^{n-1}\left(f\circ g^{-1}\right)^{(n)}(t)\,dt\right|\leq$$

$$\frac{(g(\lambda)-g(m))^{n-\frac{1}{q}}}{(p(n-1)+1)^{\frac{1}{p}}}\left\|\left(f\circ g^{-1}\right)^{(n)}\right\|_{q,[g(m),g(\lambda)]}. \tag{8.96}$$

Next, if $g(\lambda) \leq g(m)$, then

$$|f(\lambda)| = \frac{1}{(n-1)!} \left| \int_{g(\lambda)}^{g(m)} (g(\lambda) - t)^{n-1} \left(f \circ g^{-1}\right)^{(n)}(t)\, dt \right| \tag{8.97}$$

$$\leq \frac{1}{(n-1)!} \int_{g(\lambda)}^{g(m)} (t - g(\lambda))^{n-1} \left| \left(f \circ g^{-1}\right)^{(n)}(t) \right| dt$$

$$\leq \frac{1}{(n-1)!} \left(\int_{g(\lambda)}^{g(m)} (t - g(\lambda))^{p(n-1)} dt \right)^{\frac{1}{p}} \left(\int_{g(\lambda)}^{g(m)} \left| \left(f \circ g^{-1}\right)^{(n)}(t) \right|^{q} dt \right)^{\frac{1}{q}} \tag{8.98}$$

$$= \frac{1}{(n-1)!} \frac{(g(m) - g(\lambda))^{n-\frac{1}{q}}}{(p(n-1)+1)^{\frac{1}{p}}} \left\| \left(f \circ g^{-1}\right)^{(n)} \right\|_{q,[g(\lambda), g(m)]}.$$

We have proved that (if $g(\lambda) \leq g(m)$)

$$\left| \int_{g(m)}^{g(\lambda)} (g(\lambda) - t)^{n-1} \left(f \circ g^{-1}\right)^{(n)}(t)\, dt \right| \leq$$

$$\frac{(g(m) - g(\lambda))^{n-\frac{1}{q}}}{(p(n-1)+1)^{\frac{1}{p}}} \left\| \left(f \circ g^{-1}\right)^{(n)} \right\|_{q,[g(\lambda), g(m)]}. \tag{8.99}$$

Conclusion: it holds

$$\left| \int_{g(m)}^{g(\lambda)} (g(\lambda) - t)^{n-1} \left(f \circ g^{-1}\right)^{(n)}(t)\, dt \right| \leq$$

$$\frac{|g(m) - g(\lambda)|^{n-\frac{1}{q}}}{(p(n-1)+1)^{\frac{1}{p}}} \left\| \left(f \circ g^{-1}\right)^{(n)} \right\|_{q,g([m,M])}. \tag{8.100}$$

$\forall\, \lambda \in [m, M]$.

By Note 8.3, we have

$$f(A) = \frac{1}{(n-1)!} \int_{m-0}^{M} \left(\int_{g(m)}^{g(\lambda)} (g(\lambda) - t)^{n-1} \left(f \circ g^{-1}\right)^{(n)}(t)\, dt \right) dE_{\lambda}, \tag{8.101}$$

which means

$$\langle f(A)x, x \rangle = \frac{1}{(n-1)!} \int_{m-0}^{M} \left(\int_{g(m)}^{g(\lambda)} (g(\lambda) - t)^{n-1} \left(f \circ g^{-1}\right)^{(n)}(t)\, dt \right) d\langle E_{\lambda}x, x \rangle, \tag{8.102}$$

$\forall\, x \in H$.

It is well known that [11] $g_x(\lambda) := \langle E_\lambda x, x \rangle$ is nondecreasing and right continuous in λ on $[m, M]$.

Therefore it holds

$$|\langle f(A)x, x\rangle| \overset{(8.102)}{\leq} \frac{1}{(n-1)!} \int_{m-0}^{M} \left| \int_{g(m)}^{g(\lambda)} (g(\lambda) - t)^{n-1} \left(f \circ g^{-1}\right)^{(n)}(t)\, dt \right| d\langle E_\lambda x, x\rangle$$

$$\overset{(8.100)}{\leq} \frac{1}{(n-1)!} \int_{m-0}^{M} \frac{|g(m) - g(\lambda)|^{n-\frac{1}{q}}}{(p(n-1)+1)^{\frac{1}{p}}} \left\| \left(f \circ g^{-1}\right)^{(n)} \right\|_{q, g([m,M])} d\langle E_\lambda x, x\rangle \tag{8.103}$$

$$= \frac{\left\| \left(f \circ g^{-1}\right)^{(n)} \right\|_{q, g([m,M])}}{(n-1)!\,(p(n-1)+1)^{\frac{1}{p}}} \left(\int_{m-0}^{M} |g(m) - g(\lambda)|^{n-\frac{1}{q}} d\langle E_\lambda x, x\rangle \right)$$

$$= \frac{\left\| \left(f \circ g^{-1}\right)^{(n)} \right\|_{q, g([m,M])}}{(n-1)!\,(p(n-1)+1)^{\frac{1}{p}}} \left\langle |g(m) 1_H - g(A)|^{n-\frac{1}{q}} x, x \right\rangle, \tag{8.104}$$

$\forall\, x \in H$.

We have proved

Theorem 8.43 *Here all as in Theorem 8.2, with $f^{(k)}(m) = 0$, $k = 0, 1, \ldots, n-1$. Then*

$$|\langle f(A)x, x\rangle| \leq \frac{\left\| \left(f \circ g^{-1}\right)^{(n)} \right\|_{q, g([m,M])}}{(n-1)!\,(p(n-1)+1)^{\frac{1}{p}}} \left\langle |g(m) 1_H - g(A)|^{n-\frac{1}{q}} x, x \right\rangle, \tag{8.105}$$

$\forall\, x \in H$.

Inequality (8.105) means that

$$\|f(A)\| \leq \frac{\left\| \left(f \circ g^{-1}\right)^{(n)} \right\|_{q, g([m,M])}}{(n-1)!\,(p(n-1)+1)^{\frac{1}{p}}} \left\| |g(m) 1_H - g(A)|^{n-\frac{1}{q}} \right\|, \tag{8.106}$$

and in particular,

$$f(A) \leq \frac{\left\| \left(f \circ g^{-1}\right)^{(n)} \right\|_{q, g([m,M])}}{(n-1)!\,(p(n-1)+1)^{\frac{1}{p}}} |g(m) 1_H - g(A)|^{n-\frac{1}{q}}. \tag{8.107}$$

Remark 8.44 (to Theorems 8.5 and 8.6) Let $\alpha > 0$, $\lceil \alpha \rceil = n$, $\alpha \notin \mathbb{N}$.

If $\left(f \circ g^{-1}\right)^{(k)} (g(m)) = 0$, for $k = 0, 1, ..., n-1$, then

$$f(\lambda) = \frac{1}{\Gamma(\alpha)} \int_{g(m)}^{g(\lambda)} (g(\lambda) - z)^{\alpha-1} \left(\left(D_{m+;g}^{\alpha} f\right) \circ g^{-1}\right)(z) \, dz, \tag{8.108}$$

$\forall \lambda \in [m, M]$.

Hence we have

$$|f(\lambda)| \leq \frac{1}{\Gamma(\alpha)} \int_{g(m)}^{g(\lambda)} (g(\lambda) - z)^{\alpha-1} \left|\left(\left(D_{m+;g}^{\alpha} f\right) \circ g^{-1}\right)(z)\right| dz \leq$$

$$\frac{1}{\Gamma(\alpha)} \left(\int_{g(m)}^{g(\lambda)} (g(\lambda) - z)^{p(\alpha-1)} dz\right)^{\frac{1}{p}} \left(\int_{g(m)}^{g(\lambda)} \left|\left(\left(D_{m+;g}^{\alpha} f\right) \circ g^{-1}\right)(z)\right|^{q} dz\right)^{\frac{1}{q}} \leq$$

$$\frac{1}{\Gamma(\alpha)} \frac{(g(\lambda) - g(m))^{\frac{(p(\alpha-1)+1)}{p}}}{(p(\alpha-1)+1)^{\frac{1}{p}}} \left\|\left(D_{m+;g}^{\alpha} f\right) \circ g^{-1}\right\|_{q,[g(m),g(M)]}. \tag{8.109}$$

We have proved that

$$\left|\int_{g(m)}^{g(\lambda)} (g(\lambda) - z)^{\alpha-1} \left(\left(D_{m+;g}^{\alpha} f\right) \circ g^{-1}\right)(z) \, dz\right| \leq \tag{8.110}$$

$$\frac{(g(\lambda) - g(m))^{\alpha - \frac{1}{q}}}{(p(\alpha-1)+1)^{\frac{1}{p}}} \left\|\left(D_{m+;g}^{\alpha} f\right) \circ g^{-1}\right\|_{q,[g(m),g(M)]},$$

$\forall \lambda \in [m, M]$, with $\alpha > \frac{1}{q}$.

By Note 8.7 we have

$$f(A) = \frac{1}{\Gamma(\alpha)} \int_{m-0}^{M} \left(\int_{g(m)}^{g(\lambda)} (g(\lambda) - z)^{\alpha-1} \left(\left(D_{m+;g}^{\alpha} f\right) \circ g^{-1}\right)(z) \, dz\right) dE_{\lambda}, \tag{8.111}$$

which means

$$\langle f(A) x, x \rangle =$$

$$\frac{1}{\Gamma(\alpha)} \int_{m-0}^{M} \left(\int_{g(m)}^{g(\lambda)} (g(\lambda) - z)^{\alpha-1} \left(\left(D_{m+;g}^{\alpha} f\right) \circ g^{-1}\right)(z) \, dz\right) d\langle E_{\lambda} x, x \rangle, \tag{8.112}$$

$\forall x \in H$.

Therefore it holds

$$|\langle f(A)x,x\rangle| \overset{(8.112)}{\leq}$$

$$\frac{1}{\Gamma(\alpha)}\int_{m-0}^{M}\left|\int_{g(m)}^{g(\lambda)}(g(\lambda)-z)^{\alpha-1}\left(\left(D_{m+;g}^{\alpha}f\right)\circ g^{-1}\right)(z)\,dz\right| d\langle E_{\lambda}x,x\rangle \overset{(8.110)}{\leq}$$

$$\frac{1}{\Gamma(\alpha)}\int_{m-0}^{M}\frac{(g(\lambda)-g(m))^{\alpha-\frac{1}{q}}}{(p(\alpha-1)+1)^{\frac{1}{p}}}\left\|\left(D_{m+;g}^{\alpha}f\right)\circ g^{-1}\right\|_{q,[g(m),g(M)]} d\langle E_{\lambda}x,x\rangle =$$

$$\frac{\left\|\left(D_{m+;g}^{\alpha}f\right)\circ g^{-1}\right\|_{q,[g(m),g(M)]}}{\Gamma(\alpha)(p(\alpha-1)+1)^{\frac{1}{p}}}\int_{m-0}^{M}(g(\lambda)-g(m))^{\alpha-\frac{1}{q}}\,d\langle E_{\lambda}x,x\rangle =$$

$$\frac{\left\|\left(D_{m+;g}^{\alpha}f\right)\circ g^{-1}\right\|_{q,[g(m),g(M)]}}{\Gamma(\alpha)(p(\alpha-1)+1)^{\frac{1}{p}}}\langle(g(A)-g(m)1_H)^{\alpha-\frac{1}{q}}x,x\rangle, \tag{8.113}$$

$\forall\, x\in H$.

We have proved

Theorem 8.45 *Here all as in Theorem 8.6, with* $\left(f\circ g^{-1}\right)^{(k)}(g(m))=0$, *for* $k=0,1,...,n-1$, $\alpha>\frac{1}{q}$, $\alpha\notin\mathbb{N}$. *Then*

$$|\langle f(A)x,x\rangle| \leq \frac{\left\|\left(D_{m+;g}^{\alpha}f\right)\circ g^{-1}\right\|_{q,[g(m),g(M)]}}{\Gamma(\alpha)(p(\alpha-1)+1)^{\frac{1}{p}}}\left\langle(g(A)-g(m)1_H)^{\alpha-\frac{1}{q}}x,x\right\rangle, \tag{8.114}$$

$\forall\, x\in H$.

Inequality (8.114) means that

$$\|f(A)\| \leq \frac{\left\|\left(D_{m+;g}^{\alpha}f\right)\circ g^{-1}\right\|_{q,[g(m),g(M)]}}{\Gamma(\alpha)(p(\alpha-1)+1)^{\frac{1}{p}}}\left\|(g(A)-g(m)1_H)^{\alpha-\frac{1}{q}}\right\|, \tag{8.115}$$

and in particular

$$f(A) \leq \frac{\left\|\left(D_{m+;g}^{\alpha}f\right)\circ g^{-1}\right\|_{q,[g(m),g(M)]}}{\Gamma(\alpha)(p(\alpha-1)+1)^{\frac{1}{p}}}(g(A)-g(m)1_H)^{\alpha-\frac{1}{q}}. \tag{8.116}$$

We also present

Theorem 8.46 *Here all as in Theorem 8.10, with* $\left(f \circ g^{-1}\right)^{(k)}(g(M)) = 0$, *for* $k = 0, 1, ..., n-1$, $\alpha > \frac{1}{q}$, $\alpha \notin \mathbb{N}$. *Then*

$$|\langle f(A)x, x\rangle| \leq \frac{\left\| \left(D^{\alpha}_{M-;g} f\right) \circ g^{-1} \right\|_{q,[g(m),g(M)]}}{\Gamma(\alpha)(p(\alpha-1)+1)^{\frac{1}{p}}} \left\langle (g(M)1_H - g(A))^{\alpha-\frac{1}{q}} x, x \right\rangle, \tag{8.117}$$

$\forall x \in H$.

Inequality (8.117) means that

$$\|f(A)\| \leq \frac{\left\| \left(D^{\alpha}_{M-;g} f\right) \circ g^{-1} \right\|_{q,[g(m),g(M)]}}{\Gamma(\alpha)(p(\alpha-1)+1)^{\frac{1}{p}}} \left\| (g(M)1_H - g(A))^{\alpha-\frac{1}{q}} \right\|, \tag{8.118}$$

and in particular

$$f(A) \leq \frac{\left\| \left(D^{\alpha}_{M-;g} f\right) \circ g^{-1} \right\|_{q,[g(m),g(M)]}}{\Gamma(\alpha)(p(\alpha-1)+1)^{\frac{1}{p}}} (g(M)1_H - g(A))^{\alpha-\frac{1}{q}}. \tag{8.119}$$

Proof Similar to Theorem 8.45. ■

We give

Theorem 8.47 *Here all as in Theorem 8.14.*

If $\nu \geq 1$, *we assume that* $\left(f \circ g^{-1}\right)^{(k)}(g(m)) = 0$, $k = 0, 1, ..., [\nu]-1$, *and always* $\nu > \frac{1}{q}$. *Then*

$$|\langle f(A)x, x\rangle| \leq \frac{\left\| D^{\nu}_{g(m)} \left(f \circ g^{-1}\right) \right\|_{q,[g(m),g(M)]}}{\Gamma(\nu)(p(\nu-1)+1)^{\frac{1}{p}}} \left\langle (g(A) - g(m)1_H)^{\nu-\frac{1}{q}} x, x \right\rangle, \tag{8.120}$$

$\forall x \in H$.

Inequality (8.120) means that

$$\|f(A)\| \leq \frac{\left\| D^{\nu}_{g(m)} \left(f \circ g^{-1}\right) \right\|_{q,[g(m),g(M)]}}{\Gamma(\nu)(p(\nu-1)+1)^{\frac{1}{p}}} \left\| (g(A) - g(m)1_H)^{\nu-\frac{1}{q}} \right\|, \tag{8.121}$$

and in particular

$$f(A) \leq \frac{\left\| D^{\nu}_{g(m)} \left(f \circ g^{-1}\right) \right\|_{q,[g(m),g(M)]}}{\Gamma(\nu)(p(\nu-1)+1)^{\frac{1}{p}}} (g(A) - g(m)1_H)^{\nu-\frac{1}{q}}. \tag{8.122}$$

Proof Similar to Theorem 8.45. ■

We give

Theorem 8.48 *Here all as in Theorem 8.18.*

If $\nu \geq 1$, *we assume that* $\left(f \circ g^{-1}\right)^{(k)}(g(M)) = 0$, $k = 0, 1, ..., [\nu] - 1$, *and always* $\nu > \frac{1}{q}$. *Then*

$$|\langle f(A)x, x\rangle| \leq \frac{\left\|D^{\nu}_{g(M)-}\left(f \circ g^{-1}\right)\right\|_{q,[g(m),g(M)]}}{\Gamma(\nu)(p(\nu-1)+1)^{\frac{1}{p}}}\left\langle (g(M)1_H - g(A))^{\nu-\frac{1}{q}}x, x\right\rangle, \tag{8.123}$$

$\forall x \in H$.

Inequality (8.123) means that

$$\|f(A)\| \leq \frac{\left\|D^{\nu}_{g(M)-}\left(f \circ g^{-1}\right)\right\|_{q,[g(m),g(M)]}}{\Gamma(\nu)(p(\nu-1)+1)^{\frac{1}{p}}}\left\|(g(M)1_H - g(A))^{\nu-\frac{1}{q}}\right\|, \tag{8.124}$$

and in particular

$$f(A) \leq \frac{\left\|D^{\nu}_{g(M)-}\left(f \circ g^{-1}\right)\right\|_{q,[g(m),g(M)]}}{\Gamma(\nu)(p(\nu-1)+1)^{\frac{1}{p}}}(g(M)1_H - g(A))^{\nu-\frac{1}{q}}. \tag{8.125}$$

Proof Similar to Theorem 8.45. ■

We make

Remark 8.49 (to Theorem 8.21) Assume $\left(D^{i\alpha}_{*m}f\right)(m) = 0$, $i = 0, 1, ..., n$, then

$$f(\lambda) = \frac{1}{\Gamma((n+1)\alpha)}\int_m^{\lambda}(\lambda - t)^{(n+1)\alpha-1}\left(D^{(n+1)\alpha}_{*m}f\right)(t)\,dt, \tag{8.126}$$

$\forall \lambda \in [m, M]$.

We obtain

$$|f(\lambda)| \leq \frac{1}{\Gamma((n+1)\alpha)}\int_m^{\lambda}(\lambda - t)^{(n+1)\alpha-1}\left|\left(D^{(n+1)\alpha}_{*m}f\right)(t)\right|dt \leq \tag{8.127}$$

$$\frac{1}{\Gamma((n+1)\alpha)}\left(\int_m^{\lambda}(\lambda - t)^{p((n+1)\alpha-1)}dt\right)^{\frac{1}{p}}\left(\int_m^{\lambda}\left|\left(D^{(n+1)\alpha}_{*m}f\right)(t)\right|^q dt\right)^{\frac{1}{q}} \leq$$

$$\frac{1}{\Gamma((n+1)\alpha)}\frac{(\lambda - m)^{\frac{(p((n+1)\alpha-1)+1)}{p}}}{(p((n+1)\alpha-1)+1)^{\frac{1}{p}}}\left\|D^{(n+1)\alpha}_{*m}f\right\|_{q,[m,M]}. \tag{8.128}$$

We have proved that

$$\left|\int_{m}^{\lambda}(\lambda-t)^{(n+1)\alpha-1}\left(D_{*m}^{(n+1)\alpha}f\right)(t)\,dt\right|\leq$$

$$\frac{(\lambda-m)^{(n+1)\alpha-\frac{1}{q}}}{(p((n+1)\alpha-1)+1)^{\frac{1}{p}}}\left\|D_{*m}^{(n+1)\alpha}f\right\|_{q,[m,M]}, \tag{8.129}$$

$\forall\,\lambda\in[m,M]$, under $\alpha>\frac{1}{q(n+1)}$.

By Note 8.22 we have

$$\langle f(A)x,x\rangle=\frac{1}{\Gamma((n+1)\alpha)}\cdot$$

$$\int_{m-0}^{M}\left(\int_{m}^{\lambda}(\lambda-t)^{(n+1)\alpha-1}\left(D_{*m}^{(n+1)\alpha}f\right)(t)\,dt\right)d\langle E_{\lambda}x,x\rangle, \tag{8.130}$$

$\forall\,x\in H$.

Therefore

$$\langle f(A)x,x\rangle\overset{(8.129)}{\leq}$$

$$\frac{1}{\Gamma((n+1)\alpha)}\int_{m-0}^{M}\frac{\left\|D_{*m}^{(n+1)\alpha}f\right\|_{q,[m,M]}}{(p((n+1)\alpha-1)+1)^{\frac{1}{p}}}(\lambda-m)^{(n+1)\alpha-\frac{1}{q}}\,d\langle E_{\lambda}x,x\rangle=$$

$$\frac{\left\|D_{*m}^{(n+1)\alpha}f\right\|_{q,[m,M]}}{\Gamma((n+1)\alpha)(p((n+1)\alpha-1)+1)^{\frac{1}{p}}}\left\langle(A-m1_{H})^{(n+1)\alpha-\frac{1}{q}}x,x\right\rangle, \tag{8.131}$$

$\forall\,x\in H$.

We have proved:

Theorem 8.50 *Here all as in Theorem 8.21, with* $\left(D_{*m}^{i\alpha}f\right)(m)=0$, $i=0,1,...,n$. *Then*

$$|\langle f(A)x,x\rangle|\leq\frac{\left\|D_{*m}^{(n+1)\alpha}f\right\|_{q,[m,M]}}{\Gamma((n+1)\alpha)(p((n+1)\alpha-1)+1)^{\frac{1}{p}}}\left\langle(A-m1_{H})^{(n+1)\alpha-\frac{1}{q}}x,x\right\rangle, \tag{8.132}$$

$\forall\,x\in H$.

Inequality (8.132) means that

$$\|f(A)\| \leq \frac{\left\|D_{*m}^{(n+1)\alpha} f\right\|_{q,[m,M]}}{\Gamma((n+1)\alpha)(p((n+1)\alpha-1)+1)^{\frac{1}{p}}} \left\|(A - m1_H)^{(n+1)\alpha-\frac{1}{q}}\right\|, \tag{8.133}$$

and in particular,

$$f(A) \leq \frac{\left\|D_{*m}^{(n+1)\alpha} f\right\|_{q,[m,M]}}{\Gamma((n+1)\alpha)(p((n+1)\alpha-1)+1)^{\frac{1}{p}}} (A - m1_H)^{(n+1)\alpha-\frac{1}{q}}, \tag{8.134}$$

all inequalities here under $\alpha > \frac{1}{(n+1)q}$.

It follows

Theorem 8.51 *Here all as in Theorem 8.25, with* $\left(D_{M-}^{i\alpha} f\right)(M) = 0$, $i = 0, 1, ..., n$. *Then*

$$|\langle f(A)x, x\rangle| \leq \frac{\left\|D_{M-}^{(n+1)\alpha} f\right\|_{q,[m,M]}}{\Gamma((n+1)\alpha)(p((n+1)\alpha-1)+1)^{\frac{1}{p}}} \left\langle (M1_H - A)^{(n+1)\alpha-\frac{1}{q}} x, x\right\rangle, \tag{8.135}$$

$\forall\, x \in H$.

Inequality (8.135) means that

$$\|f(A)\| \leq \frac{\left\|D_{M-}^{(n+1)\alpha} f\right\|_{q,[m,M]}}{\Gamma((n+1)\alpha)(p((n+1)\alpha-1)+1)^{\frac{1}{p}}} \left\|(M1_H - A)^{(n+1)\alpha-\frac{1}{q}}\right\|, \tag{8.136}$$

and in particular,

$$f(A) \leq \frac{\left\|D_{M-}^{(n+1)\alpha} f\right\|_{q,[m,M]}}{\Gamma((n+1)\alpha)(p((n+1)\alpha-1)+1)^{\frac{1}{p}}} (M1_H - A)^{(n+1)\alpha-\frac{1}{q}}, \tag{8.137}$$

all here under $\alpha > \frac{1}{(n+1)q}$.

Proof As in Theorem 8.50. ■

We make

Remark 8.52 (to Theorem 8.29) Assume $\left(D_{m+;g}^{i\alpha} f\right)(m) = 0$, $i = 0, 1, ..., n$, then we have that

$$f(\lambda) = \frac{1}{\Gamma((n+1)\alpha)} \int_m^\lambda (g(\lambda) - g(t))^{(n+1)\alpha - 1} g'(t) \left(D_{m+;g}^{(n+1)\alpha} f\right)(t)\, dt \overset{\text{(by [14])}}{=}$$

$$\frac{1}{\Gamma((n+1)\alpha)} \int_{g(m)}^{g(\lambda)} (g(\lambda) - z)^{(n+1)\alpha - 1} \left(\left(D_{m+;g}^{(n+1)\alpha} f\right) \circ g^{-1}\right)(z)\, dz, \quad (8.138)$$

$\forall \lambda \in [m, M]$.

Hence it holds (with $\alpha > \frac{1}{q(n+1)}$)

$$|f(\lambda)| \leq \frac{1}{\Gamma((n+1)\alpha)} \left(\int_{g(m)}^{g(\lambda)} (g(\lambda) - z)^{p((n+1)\alpha - 1)}\, dz\right)^{\frac{1}{p}} \cdot$$

$$\left(\int_{g(m)}^{g(\lambda)} \left|\left(\left(D_{m+;g}^{(n+1)\alpha} f\right) \circ g^{-1}\right)(z)\right|^q dz\right)^{\frac{1}{q}} \leq$$

$$\frac{1}{\Gamma((n+1)\alpha)} \frac{(g(\lambda) - g(m))^{(n+1)\alpha - \frac{1}{q}}}{(p((n+1)\alpha - 1) + 1)^{\frac{1}{p}}} \left\|\left(D_{m+;g}^{(n+1)\alpha} f\right) \circ g^{-1}\right\|_{q,[g(m),g(M)]}, \quad (8.139)$$

$\forall \lambda \in [m, M]$.

By Note 8.30, we have

$$\langle f(A)x, x\rangle = \frac{1}{\Gamma((n+1)\alpha)} \cdot$$

$$\int_{m-0}^{M} \left(\int_{g(m)}^{g(\lambda)} (g(\lambda) - z)^{(n+1)\alpha - 1} \left(\left(D_{m+;g}^{(n+1)\alpha} f\right) \circ g^{-1}\right)(z)\, dz\right) d\langle E_\lambda x, x\rangle, \quad (8.140)$$

$\forall x \in H$.

Therefore we get

$$|\langle f(A)x, x\rangle| \overset{(8.139)}{\leq} \frac{1}{\Gamma((n+1)\alpha)} \cdot$$

$$\int_{m-0}^{M} \left[\frac{(g(\lambda) - g(m))^{(n+1)\alpha - \frac{1}{q}}}{(p((n+1)\alpha - 1) + 1)^{\frac{1}{p}}} \left\|\left(D_{m+;g}^{(n+1)\alpha} f\right) \circ g^{-1}\right\|_{q,[g(m),g(M)]}\right] d\langle E_\lambda x, x\rangle \quad (8.141)$$

$$= \frac{\left\|\left(D_{m+;g}^{(n+1)\alpha} f\right) \circ g^{-1}\right\|_{q,[g(m),g(M)]}}{\Gamma((n+1)\alpha)(p((n+1)\alpha - 1) + 1)^{\frac{1}{p}}} \left\langle (g(A) - g(m)1_H)^{(n+1)\alpha - \frac{1}{q}} x, x\right\rangle,$$

$\forall x \in H$.

We have proved:

Theorem 8.53 *Here all as in Theorem 8.29, with* $\left(D_{m+;g}^{i\alpha}f\right)(m)=0$, $i=0,1,...,n$. *Then*

$$|\langle f(A)x,x\rangle|\leq$$

$$\frac{\left\|\left(D_{m+;g}^{(n+1)\alpha}f\right)\circ g^{-1}\right\|_{q,[g(m),g(M)]}}{\Gamma((n+1)\alpha)(p((n+1)\alpha-1)+1)^{\frac{1}{p}}}\left\langle(g(A)-g(m)1_H)^{(n+1)\alpha-\frac{1}{q}}x,x\right\rangle, \tag{8.142}$$

$\forall x\in H$.

Inequality (8.142) means that

$$\|f(A)\|\leq\frac{\left\|\left(D_{m+;g}^{(n+1)\alpha}f\right)\circ g^{-1}\right\|_{q,[g(m),g(M)]}}{\Gamma((n+1)\alpha)(p((n+1)\alpha-1)+1)^{\frac{1}{p}}}\left\|(g(A)-g(m)1_H)^{(n+1)\alpha-\frac{1}{q}}\right\|, \tag{8.143}$$

and in particular,

$$f(A)\leq\frac{\left\|\left(D_{m+;g}^{(n+1)\alpha}f\right)\circ g^{-1}\right\|_{q,[g(m),g(M)]}}{\Gamma((n+1)\alpha)(p((n+1)\alpha-1)+1)^{\frac{1}{p}}}(g(A)-g(m)1_H)^{(n+1)\alpha-\frac{1}{q}}, \tag{8.144}$$

all inequalities here under $\alpha>\frac{1}{(n+1)q}$.

It follows

Theorem 8.54 *Here all as in Theorem 8.33, with* $\left(D_{M-;g}^{i\alpha}f\right)(M)=0$, $i=0,1,...,n$. *Then*

$$|\langle f(A)x,x\rangle|\leq$$

$$\frac{\left\|\left(D_{M-;g}^{(n+1)\alpha}f\right)\circ g^{-1}\right\|_{q,[g(m),g(M)]}}{\Gamma((n+1)\alpha)(p((n+1)\alpha-1)+1)^{\frac{1}{p}}}\left\langle(g(M)1_H-g(A))^{(n+1)\alpha-\frac{1}{q}}x,x\right\rangle, \tag{8.145}$$

$\forall x\in H$.

Inequality (8.145) means that

$$\|f(A)\|\leq\frac{\left\|\left(D_{M-;g}^{(n+1)\alpha}f\right)\circ g^{-1}\right\|_{q,[g(m),g(M)]}}{\Gamma((n+1)\alpha)(p((n+1)\alpha-1)+1)^{\frac{1}{p}}}\left\|(g(M)1_H-g(A))^{(n+1)\alpha-\frac{1}{q}}\right\|, \tag{8.146}$$

and in particular,

$$f(A) \leq \frac{\left\| \left(D_{M-;g}^{(n+1)\alpha} f \right) \circ g^{-1} \right\|_{q,[g(m),g(M)]}}{\Gamma((n+1)\alpha)(p((n+1)\alpha - 1) + 1)^{\frac{1}{p}}} (g(M) 1_H - g(A))^{(n+1)\alpha - \frac{1}{q}}, \tag{8.147}$$

all inequalities here under $\alpha > \frac{1}{(n+1)q}$.

Proof As in Theorem 8.53. ∎

We give

Theorem 8.55 *Here all as in Theorem 8.37 with* $\nu > \frac{1}{q(\overline{m}+1)}$. *Then*

$$|\langle f(A)x, x\rangle| \leq$$

$$\frac{\left\| D_{g(m)}^{(\overline{m}+1)\nu} \left(f \circ g^{-1} \right) \right\|_{q,[g(m),g(M)]}}{\Gamma((\overline{m}+1)\nu)(p((\overline{m}+1)\nu - 1) + 1)^{\frac{1}{p}}} \left\langle (g(A) - g(m) 1_H)^{(\overline{m}+1)\nu - \frac{1}{q}} x, x \right\rangle, \tag{8.148}$$

$\forall x \in H$.

Inequality (8.148) means that

$$\|f(A)\| \leq \frac{\left\| D_{g(m)}^{(\overline{m}+1)\nu} \left(f \circ g^{-1} \right) \right\|_{q,[g(m),g(M)]}}{\Gamma((\overline{m}+1)\nu)(p((\overline{m}+1)\nu - 1) + 1)^{\frac{1}{p}}} \left\| (g(A) - g(m) 1_H)^{(\overline{m}+1)\nu - \frac{1}{q}} \right\|, \tag{8.149}$$

and in particular,

$$f(A) \leq \frac{\left\| D_{g(m)}^{(\overline{m}+1)\nu} \left(f \circ g^{-1} \right) \right\|_{q,[g(m),g(M)]}}{\Gamma((\overline{m}+1)\nu)(p((\overline{m}+1)\nu - 1) + 1)^{\frac{1}{p}}} (g(A) - g(m) 1_H)^{(\overline{m}+1)\nu - \frac{1}{q}}. \tag{8.150}$$

Proof As in Theorem 8.53. ∎

It follows

Theorem 8.56 *Here all as in Theorem 8.40 with* $\nu > \frac{1}{q(\overline{m}+1)}$. *Then*

$$|\langle f(A)x, x\rangle| \leq$$

$$\frac{\left\| D_{g(M)-}^{(\overline{m}+1)\nu} \left(f \circ g^{-1} \right) \right\|_{q,[g(m),g(M)]}}{\Gamma((\overline{m}+1)\nu)(p((\overline{m}+1)\nu - 1) + 1)^{\frac{1}{p}}} \left\langle (g(M) 1_H - g(A))^{(\overline{m}+1)\nu - \frac{1}{q}} x, x \right\rangle, \tag{8.151}$$

$\forall x \in H$.

Inequality (8.151) means that

$$\|f(A)\| \leq \frac{\left\|D_{g(M)-}^{(\overline{m}+1)\nu}\left(f \circ g^{-1}\right)\right\|_{q,[g(m),g(M)]}}{\Gamma\left((\overline{m}+1)\nu\right)\left(p\left((\overline{m}+1)\nu-1\right)+1\right)^{\frac{1}{p}}}\left\|\left(g(M)1_H - g(A)\right)^{(\overline{m}+1)\nu-\frac{1}{q}}\right\|, \tag{8.152}$$

and in particular,

$$f(A) \leq \frac{\left\|D_{g(M)-}^{(\overline{m}+1)\nu}\left(f \circ g^{-1}\right)\right\|_{q,[g(m),g(M)]}}{\Gamma\left((\overline{m}+1)\nu\right)\left(p\left((\overline{m}+1)\nu-1\right)+1\right)^{\frac{1}{p}}}\left(g(M)1_H - g(A)\right)^{(\overline{m}+1)\nu-\frac{1}{q}}. \tag{8.153}$$

Proof As in Theorem 8.53. ■

We need

Definition 8.57 Let the real valued function $f \in C([m, M])$, and we consider

$$g(t) = \int_m^t f(z)\,dz, \forall\, t \in [m, M], \tag{8.154}$$

then $g \in C([m, M])$.

We denote by

$$\int_{m1_H}^{A} f := \Phi(g) = g(A). \tag{8.155}$$

We understand and write that $(r > 0)$

$$g^r(A) = \Phi\left(g^r\right) =: \left(\int_{m1_H}^{A} f\right)^r. \tag{8.156}$$

Clearly $\left(\int_{m1_H}^{A} f\right)^r$ is a self adjoint operator on H, for any $r > 0$.

Definition 8.58 Let $f : [m, M] \to \mathbb{R}$ be continuous. We consider

$$g(t) = \int_t^M f(z)\,dz, \forall\, t \in [m, M], \tag{8.157}$$

then $g \in C([m, M])$.

We denote by

$$\int_{A}^{M1_H} f := \Phi(g) = g(A). \tag{8.158}$$

We denote also

$$g^r(A) = \Phi(g^r) =: \left(\int_A^{M1_H} f\right)^r, \ r > 0. \tag{8.159}$$

Clearly $\left(\int_A^{M1_H} f\right)^r$ is a self adjoint operator on H, for any $r > 0$.

We mention a left fractional Poincaré type inequality:

Theorem 8.59 (by [8], pp. 385–386) *Let $g \in C^1([m, M])$ and strictly increasing. Suppose that $F_\rho := D_{m+;g}^{\rho\alpha} f$, for $\rho = 0, 1, ..., n+1$, $n \in \mathbb{N}$, fulfill: $F_\rho \circ g^{-1} \in AC([g(m), g(M)])$ and $(F_\rho \circ g^{-1})' \circ g \in L_\infty([m, M])$, where α as in (8.160) next.*

Assume that $\left(D_{m+;g}^{i\alpha} f\right)(m) = 0$, $i = 0, 1, ..., n$. Let $\gamma > 0$ with $\lceil \gamma \rceil = \overline{m}$, and $p, q > 1 : \frac{1}{p} + \frac{1}{q} = 1$. We further assume that

$$1 \geq \alpha > \max\left(\frac{\overline{m} + (k-1)\gamma}{n+1}, \frac{k\gamma q + 1}{(n+1)q}\right), \tag{8.160}$$

where $k \in \mathbb{N}$.

Then

$$\int_m^\lambda \left|D_{m+;g}^{k\gamma} f(t)\right|^q dt \leq \frac{(g(\lambda) - g(m))^{q((n+1)\alpha - k\gamma - 1) + \frac{q}{p}} (\lambda - m)}{(\Gamma((n+1)\alpha - k\gamma))^q (p((n+1)\alpha - k\gamma - 1) + 1)^{\frac{q}{p}}} \cdot$$

$$\left(\int_{g(m)}^{g(\lambda)} \left|\left(\left(D_{m+;g}^{(n+1)\alpha} f\right) \circ g^{-1}\right)(z)\right|^q dz\right) \tag{8.161}$$

$$= \frac{(g(\lambda) - g(m))^{q((n+1)\alpha - k\gamma - 1) + \frac{q}{p}} (\lambda - m)}{(\Gamma((n+1)\alpha - k\gamma))^q (p((n+1)\alpha - k\gamma - 1) + 1)^{\frac{q}{p}}} \cdot$$

$$\left(\int_m^\lambda \left|\left(D_{m+;g}^{(n+1)\alpha} f\right)(t)\right|^q g'(t)\, dt\right),$$

$\forall \lambda \in [m, M]$.

Here we have that $\left(D_{m+;g}^{k\gamma} f\right), \left(D_{m+;g}^{(n+1)\alpha} f\right) \in C([m, M])$.

Using (8.161) and properties (P) and (ii), we derive the operator left fractional Poincaré inequality:

Theorem 8.60 *All as in Theorem 8.59. Then*

$$\int_{m1_H}^{A} \left| D_{m+;g}^{k\gamma} f \right|^q \leq$$

$$\frac{\left(g\left(A\right)-g\left(m\right)1_H\right)^{q\left(\left(n+1\right)\alpha-k\gamma-1\right)+\frac{q}{p}}\left(A-m1_H\right)}{\left(\Gamma\left(\left(n+1\right)\alpha-k\gamma\right)\right)^q\left(p\left(\left(n+1\right)\alpha-k\gamma-1\right)+1\right)^{\frac{q}{p}}}\left(\int_{m1_H}^{A}\left|D_{m+;g}^{(n+1)\alpha} f\right|^q g'\right). \tag{8.162}$$

We mention a right fractional Poincaré type inequality:

Theorem 8.61 (by [8], p. 387) *Let $g \in C^1([m, M])$ and strictly increasing. Suppose that $F_\rho := D_{M-;g}^{\rho\alpha} f$, for $\rho = 0, 1, ..., n+1$, $n \in \mathbb{N}$, fulfill: $F_\rho \circ g^{-1} \in AC([g(m), g(M)])$ and $\left(F_\rho \circ g^{-1}\right)' \circ g \in L_\infty([m, M])$, where α as in (8.163) next. Assume that $\left(D_{M-;g}^{i\alpha} f\right)(M) = 0$, $i = 0, 1, ..., n$. Let $\gamma > 0$ with $\lceil \gamma \rceil = \overline{m}$, and $p, q > 1 :$ $\frac{1}{p} + \frac{1}{q} = 1$. We further assume that*

$$1 \geq \alpha > \max\left(\frac{\overline{m} + (k-1)\gamma}{n+1}, \frac{k\gamma q + 1}{(n+1)q}\right), \tag{8.163}$$

where $k \in \mathbb{N}$. Then

$$\int_{\lambda}^{M} \left| D_{M-;g}^{k\gamma} f(t) \right|^q dt \leq \frac{\left(g(M) - g(\lambda)\right)^{q((n+1)\alpha - k\gamma - 1) + \frac{q}{p}} (M - \lambda)}{\left(\Gamma((n+1)\alpha - k\gamma)\right)^q \left(p((n+1)\alpha - k\gamma - 1) + 1\right)^{\frac{q}{p}}} \cdot \tag{8.164}$$

$$\left(\int_{g(\lambda)}^{g(M)} \left| \left(\left(D_{M-;g}^{(n+1)\alpha} f \right) \circ g^{-1} \right)(z) \right|^q dz \right)$$

$$= \frac{\left(g(M) - g(\lambda)\right)^{q((n+1)\alpha - k\gamma - 1) + \frac{q}{p}} (M - \lambda)}{\left(\Gamma((n+1)\alpha - k\gamma)\right)^q \left(p((n+1)\alpha - k\gamma - 1) + 1\right)^{\frac{q}{p}}} \cdot$$

$$\left(\int_{\lambda}^{M} \left| \left(D_{M-;g}^{(n+1)\alpha} f \right)(t) \right|^q g'(t)\, dt \right),$$

$\forall \lambda \in [m, M]$.

Here we have that $\left(D_{M-;g}^{k\gamma} f\right), \left(D_{M-;g}^{(n+1)\alpha} f\right) \in C([m, M])$.

Using (8.164) and properties (P) and (ii), we derive the operator right fractional Poincaré inequality:

Theorem 8.62 *All as in Theorem 8.61. Then*

$$\int_A^{M1_H} \left| D_{M-;g}^{k\gamma} f \right|^q \leq$$

$$\frac{(g(M)1_H - g(A))^{q((n+1)\alpha - k\gamma - 1) + \frac{q}{p}} (M1_H - A)}{(\Gamma((n+1)\alpha - k\gamma))^q (p((n+1)\alpha - k\gamma - 1) + 1)^{\frac{q}{p}}} \left(\int_A^{M1_H} \left| D_{M-;g}^{(n+1)\alpha} f \right|^q g' \right). \tag{8.165}$$

We mention the following Sobolev type left fractional inequality:

Theorem 8.63 (by [7], pp. 493–495) *Let* $0 < \alpha < 1$, $f : [m, M] \to \mathbb{R}$ *such that* $f' \in L_\infty([m, M])$. *Assume that* $D_{*m}^{\overline{k}\alpha} f \in C([m, M])$, $\overline{k} = 0, 1, ..., n + 1$; $n \in \mathbb{N}$. *Suppose that* $\left(D_{*m}^{i\alpha} f\right)(m) = 0$, *for* $i = 0, 2, 3, ..., n$. *Let* $\gamma > 0$ *with* $\lceil \gamma \rceil = \overline{m}$, *and* $p, q > 1 : \frac{1}{p} + \frac{1}{q} = 1$. *We further assume that* ($k \in \mathbb{N}$)

$$1 > \alpha > \max\left(\frac{\overline{m} + (k-1)\gamma}{n+1}, \frac{k\gamma q + 1}{(n+1)q} \right), \tag{8.166}$$

Let $r \geq 1$. *Then*

$$\left(\int_m^\lambda \left| \left(D_{*m}^{k\gamma} f \right)(t) \right|^r dt \right)^{\frac{1}{r}} \leq \frac{1}{\Gamma((n+1)\alpha - k\gamma)} \cdot$$

$$\frac{(\lambda - m)^{(n+1)\alpha - k\gamma - \frac{1}{q} + \frac{1}{r}}}{(p((n+1)\alpha - k\gamma - 1) + 1)^{\frac{1}{p}} \left[r\left((n+1)\alpha - k\gamma - \frac{1}{q} \right) + 1 \right]^{\frac{1}{r}}} \cdot$$

$$\left(\int_m^\lambda \left| \left(D_{*m}^{(n+1)\alpha} f \right)(t) \right|^q dt \right)^{\frac{1}{q}}, \tag{8.167}$$

$\forall \lambda \in [m, M]$.

Here $\left(D_{*m}^{k\gamma} f \right) \in C([m, M])$.

Applying (8.167), using properties (P) and (ii), we get the following operator left fractional Sobolev type inequality:

Theorem 8.64 *All as in Theorem 8.63. Then*

$$\left(\int_{m1_H}^A \left| D_{*m}^{k\gamma} f \right|^r \right)^{\frac{1}{r}} \leq \frac{1}{\Gamma((n+1)\alpha - k\gamma)} \cdot$$

$$\frac{(A - m1_H)^{(n+1)\alpha - k\gamma - \frac{1}{q} + \frac{1}{r}}}{(p((n+1)\alpha - k\gamma - 1) + 1)^{\frac{1}{p}} \left[r\left((n+1)\alpha - k\gamma - \frac{1}{q} \right) + 1 \right]^{\frac{1}{r}}} \cdot$$

$$\left(\int_{m1_H}^{A} \left| D_{*m}^{(n+1)\alpha} f \right|^q \right)^{\frac{1}{q}}. \tag{8.168}$$

We mention the following Sobolev type right fractional inequality:

Theorem 8.65 (by [7], p. 496) *Let* $0 < \alpha < 1$, $f : [m, M] \to \mathbb{R}$ *such that* $f' \in L_\infty([m, M])$. *Assume that* $D_{M-}^{\overline{k}\alpha} f \in C([m, M])$, $\overline{k} = 0, 1, ..., n+1$; $n \in \mathbb{N}$. *Suppose that* $\left(D_{M-}^{i\alpha} f\right)(M) = 0$, *for* $i = 0, 2, 3, ..., n$. *Let* $\gamma > 0$ *with* $\lceil \gamma \rceil = \overline{m}$, *and* $p, q > 1 : \frac{1}{p} + \frac{1}{q} = 1$. *We further assume that* ($k \in \mathbb{N}$)

$$1 > \alpha > \max \left(\frac{\overline{m} + (k-1)\gamma}{n+1}, \frac{k\gamma q + 1}{(n+1)q} \right), \tag{8.169}$$

Let $r \geq 1$. *Then*

$$\left(\int_{\lambda}^{M} \left| \left(D_{M-}^{k\gamma} f \right)(t) \right|^r dt \right)^{\frac{1}{r}} \leq \frac{1}{\Gamma((n+1)\alpha - k\gamma)} \cdot$$

$$\frac{(M-\lambda)^{(n+1)\alpha - k\gamma - \frac{1}{q} + \frac{1}{r}}}{(p((n+1)\alpha - k\gamma - 1) + 1)^{\frac{1}{p}} \left[r\left((n+1)\alpha - k\gamma - \frac{1}{q} \right) + 1 \right]^{\frac{1}{r}}} \cdot \tag{8.170}$$

$$\left(\int_{\lambda}^{M} \left| \left(D_{M-}^{(n+1)\alpha} f \right)(t) \right|^q dt \right)^{\frac{1}{q}},$$

$\forall \lambda \in [m, M]$.

Here $\left(D_{M-}^{k\gamma} f \right) \in C([m, M])$.

Applying (8.170), using properties (P) and (ii), we get the following operator right fractional Sobolev type inequality:

Theorem 8.66 *All as in Theorem 8.65. Then*

$$\left(\int_{A}^{M1_H} \left| D_{M-}^{k\gamma} f \right|^r \right)^{\frac{1}{r}} \leq \frac{1}{\Gamma((n+1)\alpha - k\gamma)} \cdot$$

$$\frac{(M1_H - A)^{(n+1)\alpha - k\gamma - \frac{1}{q} + \frac{1}{r}}}{(p((n+1)\alpha - k\gamma - 1) + 1)^{\frac{1}{p}} \left[r\left((n+1)\alpha - k\gamma - \frac{1}{q} \right) + 1 \right]^{\frac{1}{r}}} \cdot$$

$$\left(\int_{A}^{M1_H} \left| D_{M-}^{(n+1)\alpha} f \right|^q \right)^{\frac{1}{q}}. \tag{8.171}$$

We need the following left fractional Poincaré type inequality:

Theorem 8.67 *Let $\alpha > 0$, $\lceil \alpha \rceil = n$, $\alpha \notin \mathbb{N}$. Let $p, q > 1 : \frac{1}{p} + \frac{1}{q} = 1$, with $\alpha > \frac{1}{q}$, and g be strictly increasing with $g \in C^1([m, M])$. We assume that $(f \circ g^{-1}) \in AC^n([g(m), g(M)])$ and $(f \circ g^{-1})^{(n)} \circ g \in L_\infty([m, M])$, and $(f \circ g^{-1})^{(k)}(g(m)) = 0$, $k = 0, 1, ..., n-1$. Then*

$$\int_m^\lambda |f(t)|^q \, dt \leq \frac{\|g\|_{\infty,[m,M]}^{q\alpha-1} (\lambda - m)^{q\alpha}}{(\Gamma(\alpha))^q (p(\alpha-1)+1)^{q-1} q\alpha} \left(\int_m^\lambda \left| \left(D_{m+;g}^\alpha f \right)(t) \right|^q g'(t) \, dt \right), \tag{8.172}$$

$\forall \lambda \in [m, M]$.

Proof By Theorem 8.5, since $(f \circ g^{-1})^{(k)}(g(m)) = 0$, for $k = 0, 1, ..., n-1$, we get that

$$f(\lambda_1) = \frac{1}{\Gamma(\alpha)} \int_{g(m)}^{g(\lambda_1)} (g(\lambda_1) - z)^{\alpha-1} \left(\left(D_{m+;g}^\alpha f \right) \circ g^{-1} \right)(z) \, dz, \tag{8.173}$$

$\forall \lambda_1 \in [m, M]$. Let $m \leq \lambda_1 \leq \lambda \leq M$. Hence we have

$$|f(\lambda_1)| \leq \frac{1}{\Gamma(\alpha)} \int_{g(m)}^{g(\lambda_1)} (g(\lambda_1) - z)^{\alpha-1} \left| \left(\left(D_{m+;g}^\alpha f \right) \circ g^{-1} \right)(z) \right| dz \leq$$

$$\frac{1}{\Gamma(\alpha)} \left(\int_{g(m)}^{g(\lambda_1)} (g(\lambda_1) - z)^{p(\alpha-1)} \, dz \right)^{\frac{1}{p}} \left(\int_{g(m)}^{g(\lambda_1)} \left| \left(\left(D_{m+;g}^\alpha f \right) \circ g^{-1} \right)(z) \right|^q dz \right)^{\frac{1}{q}} \leq$$

$$\frac{1}{\Gamma(\alpha)} \frac{(g(\lambda_1) - g(m))^{\frac{(p(\alpha-1)+1)}{p}}}{(p(\alpha-1)+1)^{\frac{1}{p}}} \left(\int_{g(m)}^{g(\lambda)} \left| \left(\left(D_{m+;g}^\alpha f \right) \circ g^{-1} \right)(z) \right|^q dz \right)^{\frac{1}{q}} \leq$$

$$\frac{1}{\Gamma(\alpha)} \frac{\|g\|_{\infty,[m,M]}^{\alpha-\frac{1}{q}} (\lambda_1 - m)^{\alpha-\frac{1}{q}}}{(p(\alpha-1)+1)^{\frac{1}{p}}} \left(\int_m^\lambda \left| \left(D_{m+;g}^\alpha f \right)(t) \right|^q g'(t) \, dt \right)^{\frac{1}{q}}. \tag{8.174}$$

We have proved that

$$|f(\lambda_1)|^q \leq \frac{\|g\|_{\infty,[m,M]}^{q\alpha-1} (\lambda_1 - m)^{q\alpha-1}}{(\Gamma(\alpha))^q (p(\alpha-1)+1)^{q-1}} \left(\int_m^\lambda \left| \left(D_{m+;g}^\alpha f \right)(t) \right|^q g'(t) \, dt \right). \tag{8.175}$$

Consequently, by integration of (8.175), we obtain

$$\int_m^\lambda |f(\lambda_1)|^q \, d\lambda_1 \leq \frac{\|g\|_{\infty,[m,M]}^{q\alpha-1} (\lambda - m)^{q\alpha-1}}{(\Gamma(\alpha))^q (p(\alpha-1)+1)^{q-1} q\alpha} \left(\int_m^\lambda \left| \left(D_{m+;g}^\alpha f \right)(t) \right|^q g'(t) \, dt \right), \tag{8.176}$$

proving the claim. ■

Using (8.172), and properties (P) and (ii), we obtain the following operator Poincaré type left fractional inequality:

Theorem 8.68 *All as in Theorem 8.67. Then*

$$\left(\int_{m1_H}^{A} |f|^q\right) \leq$$

$$\frac{\|g\|_{\infty,[m,M]}^{q\alpha-1}}{(\Gamma(\alpha))^q (p(\alpha-1)+1)^{q-1} q\alpha} (A - m1_H)^{q\alpha} \left(\int_{m1_H}^{A} \left|D_{m+;g}^{\alpha} f\right|^q g'\right). \quad (8.177)$$

We need the following left fractional Sobolev type inequality:

Theorem 8.69 *All as in Theorem 8.67 and $r \geq 1$. Then*

$$\left(\int_m^{\lambda} |f(t)|^r dt\right)^{\frac{1}{r}} \leq$$

$$\frac{\|g\|_{\infty,[m,M]}^{\left(\alpha-\frac{1}{q}\right)} (\lambda - m)^{\left(\alpha-\frac{1}{q}+\frac{1}{r}\right)}}{\Gamma(\alpha)(p(\alpha-1)+1)^{\frac{1}{p}} \left(r\left(\alpha-\frac{1}{q}\right)+1\right)^{\frac{1}{r}}} \left(\int_m^{\lambda} \left|\left(D_{m+;g}^{\alpha} f\right)(t)\right|^q g'(t)\, dt\right)^{\frac{1}{q}}, \quad (8.178)$$

$\forall\, \lambda \in [m, M]$.

Proof As in the proof of Theorem 8.67 we find that $(m \leq \lambda_1 \leq \lambda \leq M)$

$$|f(\lambda_1)| \leq \frac{\|g\|_{\infty,[m,M]}^{\alpha-\frac{1}{q}} (\lambda_1 - m)^{\alpha-\frac{1}{q}}}{\Gamma(\alpha)(p(\alpha-1)+1)^{\frac{1}{p}}} \left(\int_m^{\lambda} \left|\left(D_{m+;g}^{\alpha} f\right)(t)\right|^q g'(t)\, dt\right)^{\frac{1}{q}}. \quad (8.179)$$

Hence, by $r \geq 1$, we obtain

$$|f(\lambda_1)|^r \leq \frac{\|g\|_{\infty,[m,M]}^{r\left(\alpha-\frac{1}{q}\right)} (\lambda_1 - m)^{r\left(\alpha-\frac{1}{q}\right)}}{(\Gamma(\alpha))^r (p(\alpha-1)+1)^{\frac{r}{p}}} \left(\int_m^{\lambda} \left|\left(D_{m+;g}^{\alpha} f\right)(t)\right|^q g'(t)\, dt\right)^{\frac{r}{q}}. \quad (8.180)$$

Consequently, it holds

$$\int_m^{\lambda} |f(\lambda_1)|^r d\lambda_1 \leq$$

$$\frac{\|g\|_{\infty,[m,M]}^{r\left(\alpha-\frac{1}{q}\right)} (\lambda - m)^{r\left(\alpha-\frac{1}{q}\right)+1}}{(\Gamma(\alpha))^r (p(\alpha-1)+1)^{\frac{r}{p}} \left(r\left(a-\frac{1}{q}\right)+1\right)} \left(\int_m^{\lambda} \left|\left(D_{m+;g}^{\alpha} f\right)(t)\right|^q g'(t)\, dt\right)^{\frac{r}{q}}. \quad (8.181)$$

So that proving

$$\left(\int_m^{\lambda} |f(\lambda_1)|^r d\lambda_1\right)^{\frac{1}{r}} \leq$$

$$\frac{\|g\|_{\infty,[m,M]}^{\left(\alpha-\frac{1}{q}\right)} (\lambda-m)^{\left(\alpha-\frac{1}{q}+\frac{1}{r}\right)}}{\Gamma(\alpha)(p(\alpha-1)+1)^{\frac{1}{p}}\left(r\left(a-\frac{1}{q}\right)+1\right)^{\frac{1}{r}}} \left(\int_m^{\lambda} \left|\left(D_{m+;g}^{\alpha} f\right)(t)\right|^q g'(t)\, dt\right)^{\frac{1}{q}}. \tag{8.182}$$

■

Using (8.178), and properties (P) and (ii), we obtain the following operator Sobolev type left fractional inequality:

Theorem 8.70 *All as in Theorem 8.69. Then*

$$\left(\int_{m1_H}^{A} |f|^r\right)^{\frac{1}{r}} \leq \frac{\|g\|_{\infty,[m,M]}^{\left(\alpha-\frac{1}{q}\right)}}{\Gamma(\alpha)(p(\alpha-1)+1)^{\frac{1}{p}}\left(r\left(\alpha-\frac{1}{q}\right)+1\right)^{\frac{1}{r}}}.$$

$$(A-m1_H)^{\left(\alpha-\frac{1}{q}+\frac{1}{r}\right)}\left(\int_{m1_H}^{A} \left|\left(D_{m+;g}^{\alpha} f\right)\right|^q g'\right)^{\frac{1}{q}}. \tag{8.183}$$

We need the following right fractional Poincaré type inequality:

Theorem 8.71 *Let $\alpha > 0$, $\lceil \alpha \rceil = n$, $\alpha \notin \mathbb{N}$. Let $p, q > 1 : \frac{1}{p} + \frac{1}{q} = 1$, with $\alpha > \frac{1}{q}$, and g be strictly increasing with $g \in C^1([m, M])$. We assume that $\left(f \circ g^{-1}\right) \in AC^n([g(m), g(M)])$ and $\left(f \circ g^{-1}\right)^{(n)} \circ g \in L_{\infty}([m, M])$, and $\left(f \circ g^{-1}\right)^{(k)}(g(M)) = 0$, $k = 0, 1, ..., n-1$. Then*

$$\int_{\lambda}^{M} |f(t)|^q dt \leq \frac{\|g\|_{\infty,[m,M]}^{q\alpha-1} (M-\lambda)^{q\alpha}}{(\Gamma(\alpha))^q (p(\alpha-1)+1)^{q-1} q\alpha} \left(\int_{\lambda}^{M} \left|\left(D_{M-;g}^{\alpha} f\right)(t)\right|^q g'(t)\, dt\right), \tag{8.184}$$

$\forall \lambda \in [m, M]$.

Proof Similar to Theorem 8.67. ■

We derive the following Poincaré type right fractional inequality:

Theorem 8.72 *All as in Theorem 8.71. Then*

$$\left(\int_{A}^{M1_H} |f|^q\right) \leq$$

$$\frac{\|g\|_{\infty,[m,M]}^{q\alpha-1}}{(\Gamma(\alpha))^q (p(\alpha-1)+1)^{q-1} q\alpha} (M1_H - A)^{q\alpha} \left(\int_A^{M1_H} |D_{M-;g}^{\alpha} f|^q g'\right). \quad (8.185)$$

We need the following right fractional Sobolev type inequality:

Theorem 8.73 *All as in Theorem 8.71, $r \geq 1$. Then*

$$\left(\int_\lambda^M |f(t)|^r dt\right)^{\frac{1}{r}} \leq$$

$$\frac{\|g\|_{\infty,[m,M]}^{\left(\alpha-\frac{1}{q}\right)} (M-\lambda)^{\left(\alpha-\frac{1}{q}+\frac{1}{r}\right)}}{\Gamma(\alpha)(p(\alpha-1)+1)^{\frac{1}{p}} \left(r\left(\alpha-\frac{1}{q}\right)+1\right)^{\frac{1}{r}}} \left(\int_\lambda^M \left|\left(D_{M-;g}^{\alpha} f\right)(t)\right|^q g'(t)\, dt\right)^{\frac{1}{q}}, \quad (8.186)$$

$\forall\, \lambda \in [m, M]$.

Proof Similar to Theorem 8.69. ∎

We derive the following operator Sobolev type right fractional inequality:

Theorem 8.74 *All as in Theorem 8.73. Then*

$$\left(\int_A^{M1_H} |f|^r\right)^{\frac{1}{r}} \leq \frac{\|g\|_{\infty,[m,M]}^{\left(\alpha-\frac{1}{q}\right)}}{\Gamma(\alpha)(p(\alpha-1)+1)^{\frac{1}{p}} \left(r\left(\alpha-\frac{1}{q}\right)+1\right)^{\frac{1}{r}}}\cdot$$

$$(M1_H - A)^{\left(\alpha-\frac{1}{q}+\frac{1}{r}\right)} \left(\int_A^{M1_H} \left|\left(D_{M-;g}^{\alpha} f\right)\right|^q g'\right)^{\frac{1}{q}}. \quad (8.187)$$

We give the following left fractional Poincaré type inequality:

Theorem 8.75 *Here $n \in \mathbb{N}$ with $n = [\nu]$, $\nu > 0$. Assume that $g : [m, M] \to \mathbb{R}$ is strictly increasing function, $f \in C^n([m, M])$, $g \in C^1([m, M])$, and $g^{-1} \in C^n([g(m), g(M)])$. Suppose also that $f \circ g^{-1} \in C_{g(m)}^{\nu}([g(m), g(M)])$. If $\nu \geq 1$, we assume that $f^{(k)}(m) = 0$, all $k = 0, 1, ..., [\nu] - 1$. Let $p, q > 1 : \frac{1}{p} + \frac{1}{q} = 1$, $\nu > \frac{1}{q}$. Then*

$$\int_m^\lambda |f(t)|^q dt \leq \quad (8.188)$$

$$\frac{\|g\|_{\infty,[m,M]}^{q\nu-1} (\lambda-m)^{q\nu}}{(\Gamma(\nu))^q (p(\nu-1)+1)^{q-1} q\nu} \left(\int_m^\lambda \left|\left(D_{g(m)}^{\nu}\left(f\circ g^{-1}\right)\right)(g(s))\right|^q g'(t)\, dt\right),$$

$\forall\, \lambda \in [m, M]$.

Proof Similar to Theorem 8.67. ■

We give the following operator Poincaré type left fractional inequality:

Theorem 8.76 *All as in Theorem 8.75. Then*

$$\left(\int_{m1_H}^{A} |f|^q\right) \leq$$

$$\frac{\|g\|_{\infty,[m,M]}^{q\nu-1}}{(\Gamma(\nu))^q (p(\nu-1)+1)^{q-1} q\nu} (A - m1_H)^{q\nu} \left(\int_{m1_H}^{A} \left|\left(D_{g(m)}^{\nu}\left(f \circ g^{-1}\right)\right) \circ g\right|^q g'\right). \tag{8.189}$$

We need the following left fractional Sobolev type inequality:

Theorem 8.77 *All as in Theorem 8.75 and* $r \geq 1$. *Then*

$$\left(\int_{m}^{\lambda} |f(t)|^r dt\right)^{\frac{1}{r}} \leq \frac{\|g\|_{\infty,[m,M]}^{\left(\nu-\frac{1}{q}\right)} (\lambda - m)^{\left(\nu-\frac{1}{q}+\frac{1}{r}\right)}}{\Gamma(\nu)(p(\nu-1)+1)^{\frac{1}{p}} \left(r\left(\nu-\frac{1}{q}\right)+1\right)^{\frac{1}{r}}} \cdot$$

$$\left(\int_{m}^{\lambda} \left|\left(D_{g(m)}^{\nu}\left(f \circ g^{-1}\right)\right)(g(s))\right|^q g'(t)\, dt\right)^{\frac{1}{q}}, \tag{8.190}$$

$\forall\, \lambda \in [m, M]$.

Proof Similar to Theorem 8.69. ■

We give the corresponding operator Sobolev type left fractional inequality:

Theorem 8.78 *All as in Theorem 8.77. Then*

$$\left(\int_{m1_H}^{A} |f|^r\right)^{\frac{1}{r}} \leq \frac{\|g\|_{\infty,[m,M]}^{\left(\nu-\frac{1}{q}\right)}}{\Gamma(\nu)(p(\nu-1)+1)^{\frac{1}{p}} \left(r\left(\nu-\frac{1}{q}\right)+1\right)^{\frac{1}{r}}} \cdot$$

$$(A - m1_H)^{\left(\nu-\frac{1}{q}+\frac{1}{r}\right)} \left(\int_{m1_H}^{A} \left|\left(D_{g(m)}^{\nu}\left(f \circ g^{-1}\right)\right) \circ g\right|^q g'\right)^{\frac{1}{q}}. \tag{8.191}$$

Proof Using (8.190). ■

We give the following right fractional Poincaré type inequality:

Theorem 8.79 *Here* $n \in \mathbb{N}$ *with* $n = [\nu]$, $\nu > 0$. *Assume that* $g : [m, M] \to \mathbb{R}$ *is strictly increasing function,* $f \in C^n([m, M])$, $g \in C^1([m, M])$, *and* $g^{-1} \in C^n([g(m), g(M)])$. *Suppose also that* $f \circ g^{-1} \in C_{g(M)-}^{\nu}([g(m), g(M)])$. *If* $\nu \geq 1$, *we assume that* $f^{(k)}(M) = 0$, *all* $k = 0, 1, ..., [\nu] - 1$. *Let* $p, q > 1 : \frac{1}{p} + \frac{1}{q} = 1$, $\nu > \frac{1}{q}$. *Then*

$$\int_{\lambda}^{M} |f(t)|^q \, dt \leq \frac{\|g\|_{\infty,[m,M]}^{q\nu-1} (M-\lambda)^{q\nu}}{(\Gamma(\nu))^q (p(\nu-1)+1)^{q-1} q\nu} \cdot \tag{8.192}$$

$$\left(\int_{\lambda}^{M} \left| \left(D_{g(M)-}^{\nu} \left(f \circ g^{-1} \right) \right) (g(s)) \right|^q g'(t) \, dt \right),$$

$\forall \lambda \in [m, M]$.

Proof Similar to Theorem 8.67. ∎

We give the following operator Poincaré type right fractional inequality:

Theorem 8.80 *All as in Theorem 8.79. Then*

$$\left(\int_{A}^{M1_H} |f|^q \right) \leq \frac{\|g\|_{\infty,[m,M]}^{q\nu-1}}{(\Gamma(\nu))^q (p(\nu-1)+1)^{q-1} q\nu} \cdot \tag{8.193}$$

$$(M1_H - A)^{q\nu} \left(\int_{A}^{M1_H} \left| \left(D_{g(M)-}^{\nu} \left(f \circ g^{-1} \right) \right) \circ g \right|^q g' \right).$$

We need the following right fractional Sobolev type inequality:

Theorem 8.81 *All as in Theorem 8.79,* $r \geq 1$. *Then*

$$\left(\int_{\lambda}^{M} |f(t)|^r \, dt \right)^{\frac{1}{r}} \leq \frac{\|g\|_{\infty,[m,M]}^{\left(\nu-\frac{1}{q}\right)} (M-\lambda)^{\left(\nu-\frac{1}{q}+\frac{1}{r}\right)}}{\Gamma(\nu) (p(\nu-1)+1)^{\frac{1}{p}} \left(r\left(\nu-\frac{1}{q}\right)+1 \right)^{\frac{1}{r}}} \cdot$$

$$\left(\int_{\lambda}^{M} \left| \left(D_{g(M)-}^{\nu} \left(f \circ g^{-1} \right) \right) (g(s)) \right|^q g'(t) \, dt \right)^{\frac{1}{q}}, \tag{8.194}$$

$\forall \lambda \in [m, M]$.

Proof Similar to Theorem 8.69. ∎

We give the corresponding operator Sobolev type right fractional inequality:

Theorem 8.82 *All as in Theorem 8.81. Then*

$$\left(\int_{A}^{M1_H} |f|^r \right)^{\frac{1}{r}} \leq \frac{\|g\|_{\infty,[m,M]}^{\left(\nu-\frac{1}{q}\right)}}{\Gamma(\nu) (p(\nu-1)+1)^{\frac{1}{p}} \left(r\left(\nu-\frac{1}{q}\right)+1 \right)^{\frac{1}{r}}} \cdot$$

$$(M1_H - A)^{\left(\nu - \frac{1}{q} + \frac{1}{r}\right)} \left(\int_A^{M1_H} \left| \left(D^{\nu}_{g(M)-} \left(f \circ g^{-1} \right) \right) \circ g \right|^q g' \right)^{\frac{1}{q}}. \tag{8.195}$$

Proof Using (8.194). ■

We give the following Poincaré type left fractional inequality:

Theorem 8.83 *Let $g : [m, M] \to \mathbb{R}$ be strictly increasing, $f \in C^1([m, M])$, $g \in C^1([m, M])$, and $g^{-1} \in C^1([g(m), g(M)])$, $0 < \nu < 1$. Suppose that $\left(D^{i\nu}_{g(m)} \left(f \circ g^{-1} \right) \right) \in C^{\nu}_{g(m)}([g(m), g(M)])$, $i = 0, 1, ..., \overline{m}$, and $\left(D^{(\overline{m}+1)\nu}_{g(m)} \left(f \circ g^{-1} \right) \right) \in C([g(m), g(M)])$. Let $p, q > 1 : \frac{1}{p} + \frac{1}{q} = 1$, with $\nu > \frac{1}{(\overline{m}+1)q}$. Then*

$$\int_m^{\lambda} |f(t)|^q dt \leq \frac{\|g\|^{q(\overline{m}+1)\nu - 1}_{\infty,[m,M]}}{(\Gamma((\overline{m}+1)\nu))^q (p((\overline{m}+1)\nu - 1) + 1)^{q-1} q(\overline{m}+1)\nu}. \tag{8.196}$$

$$(\lambda - m)^{q(\overline{m}+1)\nu} \left(\int_m^{\lambda} \left| \left(D^{(\overline{m}+1)\nu}_{g(m)} \left(f \circ g^{-1} \right) \right) (g(s)) \right|^q g'(t) \, dt \right),$$

$\forall \lambda \in [m, M]$.

Proof Similar to Theorem 8.67. ■

We give the corresponding operator Poincaré type left fractional inequality:

Theorem 8.84 *All as in Theorem 8.83. Then*

$$\left(\int_{m1_H}^{A} |f|^q \right) \leq \frac{\|g\|^{q(\overline{m}+1)\nu - 1}_{\infty,[m,M]}}{(\Gamma((\overline{m}+1)\nu))^q (p((\overline{m}+1)\nu - 1) + 1)^{q-1} q(\overline{m}+1)\nu}.$$

$$(A - m1_H)^{q(\overline{m}+1)\nu} \left(\int_{m1_H}^{A} \left| D^{(\overline{m}+1)\nu}_{g(m)} \left(f \circ g^{-1} \right) \circ g \right|^q g' \right). \tag{8.197}$$

We need the following left fractional Sobolev type inequality:

Theorem 8.85 *All as in Theorem 8.83, and $r \geq 1$. Then*

$$\left(\int_m^{\lambda} |f(t)|^r dt \right)^{\frac{1}{r}} \leq \tag{8.198}$$

$$\frac{\|g\|^{(\overline{m}+1)\nu - \frac{1}{q}}_{\infty,[m,M]} (\lambda - m)^{\left((\overline{m}+1)\nu - \frac{1}{q} + \frac{1}{r}\right)}}{\Gamma((\overline{m}+1)\nu) (p((\overline{m}+1)\nu - 1) + 1)^{\frac{1}{p}} \left(r\left((\overline{m}+1)\nu - \frac{1}{q}\right) + 1 \right)^{\frac{1}{r}}}.$$

$$\left(\int_m^\lambda \left|\left(D_{g(m)}^{(\overline{m}+1)\nu}\left(f\circ g^{-1}\right)\right)(g(s))\right|^q g'(t)\,dt\right)^{\frac{1}{q}},$$

$\forall\, \lambda \in [m, M]$.

Proof Similar to Theorem 8.69. ■

We give the corresponding operator Sobolev type left fractional inequality:

Theorem 8.86 *All as in Theorem 8.85. Then*

$$\left(\int_{m1_H}^A |f|^r\right)^{\frac{1}{r}} \leq$$

$$\frac{\|g\|_{\infty,[m,M]}^{(\overline{m}+1)\nu-\frac{1}{q}}}{\Gamma\left((\overline{m}+1)\nu\right)\left(p\left((\overline{m}+1)\nu-1\right)+1\right)^{\frac{1}{p}}\left(r\left((\overline{m}+1)\nu-\frac{1}{q}\right)+1\right)^{\frac{1}{r}}}.$$

$$(A-m1_H)^{\left((\overline{m}+1)\nu-\frac{1}{q}+\frac{1}{r}\right)}\left(\int_{m1_H}^A \left|\left(D_{g(m)}^{(\overline{m}+1)\nu}\left(f\circ g^{-1}\right)\right)\circ g\right|^q g'\right)^{\frac{1}{q}}. \tag{8.199}$$

Proof Using (8.198). ■

We give the following Poincaré type right fractional inequality:

Theorem 8.87 *Let $g : [m, M] \to \mathbb{R}$ be strictly increasing, $f \in C^1([m, M])$, $g \in C^1([m, M])$, and $g^{-1} \in C^1([g(m), g(M)])$, $0 < \nu < 1$. Suppose that $\left(D_{g(M)-}^{i\nu}\left(f\circ g^{-1}\right)\right) \in C_{g(M)-}^{\nu}([g(m), g(M)])$, $i = 0, 1, ..., \overline{m}$, and $\left(D_{g(M)-}^{(\overline{m}+1)\nu}\left(f\circ g^{-1}\right)\right) \in C([g(m), g(M)])$. Let $p, q > 1 : \frac{1}{p}+\frac{1}{q} = 1$, with $\nu > \frac{1}{(\overline{m}+1)q}$. Then*

$$\int_\lambda^M |f(t)|^q\,dt \leq \frac{\|g\|_{\infty,[m,M]}^{q(\overline{m}+1)\nu-1}}{\left(\Gamma\left((\overline{m}+1)\nu\right)\right)^q\left(p\left((\overline{m}+1)\nu-1\right)+1\right)^{q-1}q\left(\overline{m}+1\right)\nu}. \tag{8.200}$$

$$(M-\lambda)^{q(\overline{m}+1)\nu}\left(\int_\lambda^M \left|\left(D_{g(M)-}^{(\overline{m}+1)\nu}\left(f\circ g^{-1}\right)\right)(g(s))\right|^q g'(t)\,dt\right),$$

$\forall\, \lambda \in [m, M]$.

Proof Similar to Theorem 8.67. ■

We give the corresponding operator Poincaré type right fractional inequality:

Theorem 8.88 *All as in Theorem 8.87. Then*

$$\left(\int_A^{M1_H} |f|^q\right) \leq \frac{\|g\|_{\infty,[m,M]}^{q(\overline{m}+1)\nu-1}}{\left(\Gamma\left((\overline{m}+1)\nu\right)\right)^q\left(p\left((\overline{m}+1)\nu-1\right)+1\right)^{q-1}q\left(\overline{m}+1\right)\nu}.$$

$$(M1_H - A)^{q(\overline{m}+1)\nu} \left(\int_A^{M1_H} \left| D_{g(M)-}^{(\overline{m}+1)\nu} \left(f \circ g^{-1} \right) \circ g \right|^q g' \right). \tag{8.201}$$

We need the following right fractional Sobolev type inequality:

Theorem 8.89 *All as in Theorem 8.87, $r \geq 1$. Then*

$$\left(\int_{\lambda}^{M} |f(t)|^r dt \right)^{\frac{1}{r}} \leq$$

$$\frac{\|g\|_{\infty,[m,M]}^{(\overline{m}+1)\nu - \frac{1}{q}}}{\Gamma((\overline{m}+1)\nu)(p((\overline{m}+1)\nu - 1) + 1)^{\frac{1}{p}} \left(r\left((\overline{m}+1)\nu - \frac{1}{q} \right) + 1 \right)^{\frac{1}{r}}}.$$

$$(M - \lambda)^{\left((\overline{m}+1)\nu - \frac{1}{q} + \frac{1}{r} \right)} \left(\int_{\lambda}^{M} \left| \left(D_{g(M)-}^{(\overline{m}+1)\nu} \left(f \circ g^{-1} \right) \right) (g(s)) \right|^q g'(t)\, dt \right)^{\frac{1}{q}}, \tag{8.202}$$

$\forall \lambda \in [m, M]$.

Proof Similar to Theorem 8.69. ■

We finish with the corresponding operator Sobolev type right fractional inequality:

Theorem 8.90 *All as in Theorem 8.89. Then*

$$\left(\int_A^{M1_H} |f|^r \right)^{\frac{1}{r}} \leq$$

$$\frac{\|g\|_{\infty,[m,M]}^{(\overline{m}+1)\nu - \frac{1}{q}}}{\Gamma((\overline{m}+1)\nu)(p((\overline{m}+1)\nu - 1) + 1)^{\frac{1}{p}} \left(r\left((\overline{m}+1)\nu - \frac{1}{q} \right) + 1 \right)^{\frac{1}{r}}}.$$

$$(M1_H - A)^{\left((\overline{m}+1)\nu - \frac{1}{q} + \frac{1}{r} \right)} \left(\int_A^{M1_H} \left| \left(D_{g(M)-}^{(\overline{m}+1)\nu} \left(f \circ g^{-1} \right) \right) \circ g \right|^q g' \right)^{\frac{1}{q}}. \tag{8.203}$$

Proof Using (8.202). ■

References

1. G.A. Anastassiou, *Fractional Differentiation Inequalities* (Springer, New York, 2009)
2. G.A. Anastassiou, *Intelligent Mathematics: Computational Analysis* (Springer, Heidelberg, New York, 2011)

3. G. Anastassiou, Advanced fractional Taylor's formulae. J. Comput. Anal. Appl. **21**(7), 1185–1204 (2016)
4. G. Anastassiou, Generalized canavati type fractional Taylor's formulae. J. Comput. Anal. Appl. **21**(7), 1205–1212 (2016)
5. G. Anastassiou, *Most General Fractional Self Adjoint Operator Representation formulae and Operator Poincaré and Sobolev type and other basic Inequalities*, submitted (2016)
6. G. Anastassiou, I.K. Argyros, *A convergence Analysis for a certain family of extended iterative methods: Part II. Applications to fractional calculus*. Ann. Univ. Sci. Budapest, Sect. Comput., accepted (2015)
7. G. Anastassiou, O. Duman (eds.), *Intelligent Mathematics II: Applied Mathematics and Approximation Theory* (Springer, Heidelberg, 2016)
8. G. Anastassiou, O. Duman (eds.), *Computational Analysis, AMAT, Ankara, May 2015* (Springer, New York, 2016)
9. K. Diethelm, *The Analysis of Fractional Differential Equations* (Springer, New York, 2010)
10. S.S. Dragomir, *Inequalities for functions of selfadjoint operators on Hilbert Spaces* (2011). ajmaa.org/RGMIA/monographs/InFuncOp.pdf
11. S. Dragomir, *Operator Inequalities of Ostrowski and Trapezoidal Type* (Springer, New York, 2012)
12. T. Furuta, J. Mićić Hot, J. Pečaric, Y. Seo, *Mond-Pečaric Method in Operator Inequalities. Inequalities for Bounded Selfadjoint Operators on a Hilbert Space* (Element, Zagreb, 2005)
13. G. Helmberg, *Introduction to Spectral Theory in Hilbert Space* (Wiley, New York, 1969)
14. R.-Q. Jia, *Chapter 3. Absolutely Continuous Functions*. https://www.ualberta.ca/~rjia/Math418/Notes/Chap.3.pdf
15. Z.M. Odibat, N.J. Shawagleh, Generalized Taylor's formula. Appl. Math. Comput. **186**, 286–293 (2007)

Chapter 9
Harmonic Self Adjoint Operator Chebyshev-Grüss Type Inequalities

We present here very general self adjoint operator harmonic Chebyshev-Grü ss inequalities with applications. It follows [3].

9.1 Motivation

Here we mention the following inspiring and motivating result.

Theorem 9.1 (Čebyšev 1882, [4]). *Let $f, g : [a, b] \to \mathbb{R}$ absolutely continuous functions. If $f', g' \in L_\infty([a, b])$, then*

$$\left| \frac{1}{b-a} \int_a^b f(x) g(x) \, dx - \left(\frac{1}{b-a} \int_a^b f(x) \, dx \right) \left(\frac{1}{b-a} \int_a^b g(x) \, dx \right) \right| \tag{9.1}$$

$$\leq \frac{1}{12} (b-a)^2 \left\| f' \right\|_\infty \left\| g' \right\|_\infty .$$

Also we mention

Theorem 9.2 (Grüss 1935, [10]). *Let f, g integrable functions from $[a, b]$ into $\mathbb{R}$, such that $m \leq f(x) \leq M$, $\rho \leq g(x) \leq \sigma$, for all $x \in [a, b]$, where $m, M, \rho, \sigma \in \mathbb{R}$. Then*

$$\left| \frac{1}{b-a} \int_a^b f(x) g(x) \, dx - \left(\frac{1}{b-a} \int_a^b f(x) \, dx \right) \left(\frac{1}{b-a} \int_a^b g(x) \, dx \right) \right| \tag{9.2}$$

$$\leq \frac{1}{4} (M - m) (\sigma - \rho) .$$

Next we follow [1], pp. 132–152.

G.A. Anastassiou, *Intelligent Comparisons II: Operator Inequalities and Approximations*, Studies in Computational Intelligence 699,
DOI 10.1007/978-3-319-51475-8_9

We make

Brief Assumption 9.3 *Let* $f : \prod_{i=1}^{m} [a_i, b_i] \to \mathbb{R}$ *with* $\frac{\partial^l f}{\partial x_i^l}$ *for* $l = 0, 1, ..., n$; $i = 1, ..., m$, *are continuous on* $\prod_{i=1}^{m} [a_i, b_i]$.

Definition 9.4 *We put*

$$q(x_i, s_i) = \begin{cases} s_i - a_i, & if \; s_i \in [a_i, x_i], \\ s_i - b_i, & if \; s_i \in (x_i, b_i], \end{cases} \tag{9.3}$$

$x_i \in [a_i, b_i]$, $i = 1, ..., m$.

Let $(P_n)_{n \in \mathbb{N}}$ *be a harmonic sequence of polynomials, that is* $P_n' = P_{n-1}$, $n \in \mathbb{N}$, $P_0 = 1$.

Let functions f_λ, $\lambda = 1, ..., r \in \mathbb{N} - \{1\}$, *as in Brief Assumption 9.3, and* $n_\lambda \in \mathbb{N}$ *associated with* f_λ.

We set

$$A_{i\lambda}(x_i, ..., x_m) := \frac{n_\lambda^{i-1}}{\prod_{j=1}^{i-1} (b_j - a_j)} \cdot$$

$$\left[\sum_{k=1}^{n_\lambda - 1} (-1)^{k+1} P_k(x_i) \int_{a_1}^{b_1} ... \int_{a_{i-1}}^{b_{i-1}} \frac{\partial^k f_\lambda(s_1, ..., s_{i-1}, x_i, ..., x_m)}{\partial x_i^k} ds_1 ... ds_{i-1} + \right. \tag{9.4}$$

$$\sum_{k=1}^{n_\lambda - 1} \frac{(-1)^k (n_\lambda - k)}{b_i - a_i} \cdot$$

$$\left[P_k(b_i) \int_{a_1}^{b_1} ... \int_{a_{i-1}}^{b_{i-1}} \frac{\partial^{k-1} f_\lambda(s_1, ..., s_{i-1}, b_i, x_{i+1}, ..., x_m)}{\partial x_i^{k-1}} ds_1 ... ds_{i-1} - \right.$$

$$\left. \left. P_k(a_i) \int_{a_1}^{b_1} ... \int_{a_{i-1}}^{b_{i-1}} \frac{\partial^{k-1} f_\lambda(s_1, ..., s_{i-1}, a_i, x_{i+1}, ..., x_m)}{\partial x_i^{k-1}} ds_1 ... ds_{i-1} \right] \right],$$

and

$$B_{i\lambda}(x_i, ..., x_m) := \frac{n_\lambda^{i-1} (-1)^{n_\lambda + 1}}{\prod_{j=1}^{i} (b_j - a_j)} \cdot \tag{9.5}$$

$$\left[\int_{a_1}^{b_1} ... \int_{a_i}^{b_i} P_{n_\lambda - 1}(s_i) q(x_i, s_i) \frac{\partial^{n_\lambda} f_\lambda(s_1, ..., s_i, x_{i+1}, ..., x_m)}{\partial x_i^{n_\lambda}} ds_1 ... ds_i \right],$$

for all $i = 1, ..., m$; $\lambda = 1, ..., r$.

We also set

$$A_1 := \left(\frac{\left(\prod_{j=1}^{m} (b_j - a_j) \right)}{3} \right) \cdot \left[\sum_{\lambda=1}^{r} \left\{ \left(\prod_{\substack{\rho=1 \\ \rho \neq \lambda}}^{r} \| f_\rho \|_{\infty, \prod_{j=1}^{m} [a_j, b_j]} \right) \cdot \right. \right. \tag{9.6}$$

$$\left. \left. \left(\sum_{i=1}^{m} \left[(b_i - a_i) \, n_\lambda^{i-1} \, \| P_{n_\lambda - 1} \|_{\infty, [a_i, b_i]} \left\| \frac{\partial^{n_\lambda} f_\lambda}{\partial x_i^{n_\lambda}} \right\|_{\infty, \prod_{j=1}^{m} [a_j, b_j]} \right] \right) \right\} \right],$$

(let $p, q > 1 : \frac{1}{p} + \frac{1}{q} = 1$*)*

$$A_2 := \sum_{\lambda=1}^{r} \sum_{i=1}^{m} \left\| \prod_{\substack{\rho=1 \\ \rho \neq \lambda}}^{r} f_\rho \right\|_{L_p\left(\prod_{j=1}^{m} [a_j, b_j] \right)} \| B_{i\lambda} \|_{L_q\left(\prod_{j=i}^{m} [a_j, b_j] \right)} \left(\prod_{j=1}^{i-1} (b_j - a_j) \right)^{\frac{1}{q}}, \tag{9.7}$$

and

$$A_3 := \frac{1}{2} \left\{ \sum_{\lambda=1}^{r} \left\{ \left\| \prod_{\substack{\rho=1 \\ \rho \neq \lambda}}^{r} f_\rho \right\|_{L_1\left(\prod_{j=1}^{m} [a_j, b_j] \right)} \left[\sum_{i=1}^{m} \left[(b_i - a_i) \, n_\lambda^{i-1} \cdot \right. \right. \right. \right. \tag{9.8}$$

$$\left. \left. \left. \left. \| P_{n_\lambda - 1} \|_{\infty, [a_i, b_i]} \left\| \frac{\partial^{n_\lambda} f_\lambda}{\partial x_i^{n_\lambda}} \right\|_{\infty, \prod_{j=1}^{m} [a_j, b_j]} \right] \right] \right\} \right\}.$$

We finally set

$$W := r \int_{\prod_{j=1}^{m} [a_j, b_j]} \left(\prod_{\rho=1}^{r} f_\rho (x) \right) dx - \tag{9.9}$$

$$\frac{1}{\prod_{j=1}^{n} (b_j - a_j)} \sum_{\lambda=1}^{r} n_\lambda^m \left(\int_{\prod_{j=1}^{m} [a_j, b_j]} \left(\prod_{\substack{\rho=1 \\ \rho \neq \lambda}}^{r} f_\rho (x) \right) dx \right) \left(\int_{\prod_{j=1}^{m} [a_j, b_j]} f_\lambda (s) \, ds \right)$$

$$-\sum_{\lambda=1}^{r}\int_{\prod_{j=1}^{m}[a_j,b_j]}\left(\left(\prod_{\substack{\rho=1\\ \rho\neq\lambda}}^{r}f_\rho(x)\right)\left(\sum_{i=1}^{m}A_{i\lambda}(x_i,...,x_m)\right)\right)dx.$$

We mention

Theorem 9.5 ([1], p. 151–152). *It holds*

$$|W|\leq\min\{A_1,A_2,A_3\}. \tag{9.10}$$

9.2 Background

Let A be a selfadjoint linear operator on a complex Hilbert space $(H;\langle\cdot,\cdot\rangle)$. The Gelfand map establishes a $*-$isometrically isomorphism Φ between the set $C(Sp(A))$ of all continuous functions defined on the spectrum of A, denoted $Sp(A)$, and the C^*-algebra $C^*(A)$ generated by A and the identity operator 1_H on H as follows (see e.g. [9, p. 3]):

For any $f,g\in C(Sp(A))$ and any $\alpha,\beta\in\mathbb{C}$ we have

(i) $\Phi(\alpha f+\beta g)=\alpha\Phi(f)+\beta\Phi(g)$;

(ii) $\Phi(fg)=\Phi(f)\Phi(g)$ (the operation composition is on the right) and $\Phi(\overline{f})=(\Phi(f))^*$;

(iii) $\|\Phi(f)\|=\|f\|:=\sup_{t\in Sp(A)}|f(t)|$;

(iv) $\Phi(f_0)=1_H$ and $\Phi(f_1)=A$, where $f_0(t)=1$ and $f_1(t)=t$, for $t\in Sp(A)$.

With this notation we define

$$f(A):=\Phi(f), \text{ for all } f\in C(Sp(A)),$$

and we call it the continuous functional calculus for a selfadjoint operator A.

If A is a selfadjoint operator and f is a real valued continuous function on $Sp(A)$ then $f(t)\geq 0$ for any $t\in Sp(A)$ implies that $f(A)\geq 0$, i.e. $f(A)$ is a positive operator on H. Moreover, if both f and g are real valued functions on $Sp(A)$ then the following important property holds:

(P) $f(t)\geq g(t)$ for any $t\in Sp(A)$, implies that $f(A)\geq g(A)$ in the operator order of $B(H)$.

Equivalently, we use (see [7], pp. 7–8):

Let U be a selfadjoint operator on the complex Hilbert space $(H,\langle\cdot,\cdot\rangle)$ with the spectrum $Sp(U)$ included in the interval $[m,M]$ for some real numbers $m<M$ and $\{E_\lambda\}_\lambda$ be its spectral family.

Then for any continuous function $f : [m, M] \to \mathbb{C}$, it is well known that we have the following spectral representation in terms of the Riemann-Stieljes integral:

$$\langle f(U)x, y\rangle = \int_{m-0}^{M} f(\lambda)\, d(\langle E_\lambda x, y\rangle), \tag{9.11}$$

for any $x, y \in H$. The function $g_{x,y}(\lambda) := \langle E_\lambda x, y\rangle$ is of bounded variation on the interval $[m, M]$, and

$$g_{x,y}(m-0) = 0 \quad \text{and} \quad g_{x,y}(M) = \langle x, y\rangle,$$

for any $x, y \in H$. Furthermore, it is known that $g_x(\lambda) := \langle E_\lambda x, x\rangle$ is increasing and right continuous on $[m, M]$.

An important formula used a lot here is

$$\langle f(U)x, x\rangle = \int_{m-0}^{M} f(\lambda)\, d(\langle E_\lambda x, x\rangle), \quad \forall\, x \in H. \tag{9.12}$$

As a symbol we can write

$$f(U) = \int_{m-0}^{M} f(\lambda)\, dE_\lambda. \tag{9.13}$$

Above, $m = \min\{\lambda | \lambda \in Sp(U)\} := \min Sp(U)$, $M = \max\{\lambda | \lambda \in Sp(U)\} := \max Sp(U)$. The projections $\{E_\lambda\}_{\lambda\in\mathbb{R}}$, are called the spectral family of A, with the properties:

(a) $E_\lambda \le E_{\lambda'}$ for $\lambda \le \lambda'$;

(b) $E_{m-0} = 0_H$ (zero operator), $E_M = 1_H$ (identity operator) and $E_{\lambda+0} = E_\lambda$ for all $\lambda \in \mathbb{R}$.

Furthermore

$$E_\lambda := \varphi_\lambda(U), \quad \forall\, \lambda \in \mathbb{R}, \tag{9.14}$$

is a projection which reduces U, with

$$\varphi_\lambda(s) := \begin{cases} 1, & \text{for } -\infty < s \le \lambda, \\ 0, & \text{for } \lambda < s < +\infty. \end{cases}$$

The spectral family $\{E_\lambda\}_{\lambda\in\mathbb{R}}$ determines uniquely the self-adjoint operator U and vice versa.

For more on the topic see [11], pp. 256–266, and for more details see there pp. 157–266. See also [6].

Some more basics are given (we follow [7], pp. 1–5):

Let $(H; \langle \cdot, \cdot \rangle)$ be a Hilbert space over $\mathbb{C}$. A bounded linear operator A defined on H is selfjoint, i.e., $A = A^*$, iff $\langle Ax, x \rangle \in \mathbb{R}$, $\forall\, x \in H$, and if A is selfadjoint, then

$$\|A\| = \sup_{x \in H : \|x\| = 1} |\langle Ax, x \rangle| . \tag{9.15}$$

Let A, B be selfadjoint operators on H. Then $A \leq B$ iff $\langle Ax, x \rangle \leq \langle Bx, x \rangle$, $\forall$ $x \in H$.

In particular, A is called positive if $A \geq 0$.

Denote by

$$\mathcal{P} := \left\{ \varphi(s) := \sum_{k=0}^{n} \alpha_k s^k | n \geq 0, \alpha_k \in \mathbb{C}, 0 \leq k \leq n \right\} . \tag{9.16}$$

If $A \in \mathcal{B}(H)$ (the Banach algebra of all bounded linear operators defined on H, i.e. from H into itself) is selfadjoint, and $\varphi(s) \in \mathcal{P}$ has real coefficients, then $\varphi(A)$ is selfadjoint, and

$$\|\varphi(A)\| = \max \{ |\varphi(\lambda)| , \lambda \in Sp(A) \} . \tag{9.17}$$

If φ is any function defined on $\mathbb{R}$ we define

$$\|\varphi\|_A := \sup \{ |\varphi(\lambda)| , \lambda \in Sp(A) \} . \tag{9.18}$$

If A is selfadjoint operator on Hilbert space H and φ is continuous and given that $\varphi(A)$ is selfadjoint, then $\|\varphi(A)\| = \|\varphi\|_A$. And if φ is a continuous real valued function so it is $|\varphi|$, then $\varphi(A)$ and $|\varphi|(A) = |\varphi(A)|$ are selfadjoint operators (by [7], p. 4, Theorem 7).

Hence it holds

$$\| |\varphi(A)| \| = \| |\varphi| \|_A = \sup \{ \| \varphi(\lambda) \| , \lambda \in Sp(A) \}$$

$$= \sup \{ |\varphi(\lambda)| , \lambda \in Sp(A) \} = \|\varphi\|_A = \|\varphi(A)\| ,$$

that is

$$\| |\varphi(A)| \| = \|\varphi(A)\| . \tag{9.19}$$

For a selfadjoint operator $A \in \mathcal{B}(H)$ which is positive, there exists a unique positive selfadjoint operator $B := \sqrt{A} \in \mathcal{B}(H)$ such that $B^2 = A$, that is $\left(\sqrt{A}\right)^2 = A$. We call B the square root of A.

Let $A \in \mathcal{B}(H)$, then $A^* A$ is selfadjoint and positive. Define the "operator absolute value" $|A| := \sqrt{A^* A}$. If $A = A^*$, then $|A| = \sqrt{A^2}$.

For a continuous real valued function φ we observe the following:

$$|\varphi(A)| \text{ (the functional absolute value) } = \int_{m-0}^{M} |\varphi(\lambda)| \, dE_\lambda =$$

$$\int_{m-0}^{M} \sqrt{(\varphi(\lambda))^2} dE_\lambda = \sqrt{(\varphi(A))^2} = |\varphi(A)| \text{ (operator absolute value)},$$

where A is a selfadjoint operator.

That is we have

$$|\varphi(A)| \text{ (functional absolute value) } = |\varphi(A)| \text{ (operator absolute value)}. \quad (9.20)$$

Let $A, B \in \mathcal{B}(H)$, then

$$\|AB\| \leq \|A\| \, \|B\|, \quad (9.21)$$

by Banach algebra property.

9.3 Main Results

Let $(P_n)_{n \in \mathbb{N}}$ be a harmonic sequence of polynomials, that is $P_n' = P_{n-1}$, $n \in \mathbb{N}$, $P_0 = 1$. Furthermore, let $[a, b] \subset \mathbb{R}$, $a \neq b$, and $h : [a, b] \to \mathbb{R}$ be such that $h^{(n-1)}$ is absolutely continuous function for some $n \in \mathbb{N}$.

We set

$$q(x, t) = \begin{cases} t - a, & \text{if } t \in [a, x], \\ t - b, & \text{if } t \in (x, b], \end{cases} \quad x \in [a, b]. \quad (9.22)$$

By [5], and [1], p. 133, we get the generalized Fink type representation formula

$$h(x) = \sum_{k=1}^{n-1} (-1)^{k+1} P_k(x) h^{(k)}(x) +$$

$$\sum_{k=1}^{n-1} \frac{(-1)^k (n-k)}{b-a} \left[P_k(b) h^{(k-1)}(b) - P_k(a) h^{(k-1)}(a) \right] + \quad (9.23)$$

$$\frac{n}{b-a} \int_a^b h(t) \, dt + \frac{(-1)^{n+1}}{b-a} \int_a^b P_{n-1}(t) \, q(x, t) \, h^{(n)}(t) \, dt,$$

$\forall\, x \in [a, b]$, $n \in \mathbb{N}$, when $n = 1$ the above sums are zero.

For the harmonic sequence of polynomials $P_k(t) = \frac{(t-x)^k}{k!}$, $k \in \mathbb{Z}_+$, (9.23) reduces to Fink formula, see [8].

Next we present very general harmonic Chebyshev-Grüss operator inequalities based on (9.23). Then we specialize them for $n = 1$.

We give

Theorem 9.6 *Let $n \in \mathbb{N}$ and $f, g \in C^n([a,b])$ with $[m, M] \subset (a,b)$, $m < M$. Here A is a selfadjoint linear bounded operator on the Hilbert space H with spectrum $Sp(A) \subseteq [m, M]$. We consider any $x \in H : \|x\| = 1$.*

Then

$$\langle (\Delta(f,g))(A)x, x\rangle := |\langle f(A) g(A) x, x\rangle - \langle f(A)x, x\rangle \langle g(A)x, x\rangle -$$

$$\frac{1}{2}\left[\sum_{k=1}^{n-1} (-1)^{k+1} \left\{\left[\left\langle P_k(A)\left(g(A) f^{(k)}(A) + f(A) g^{(k)}(A)\right)x, x\right\rangle\right] - \right.\right. \tag{9.24}$$

$$\left[\left\langle P_k(A) f^{(k)}(A)x, x\right\rangle \langle g(A)x, x\rangle + \left\langle P_k(A) g^{(k)}(A)x, x\right\rangle \langle f(A)x, x\rangle\right]\}]| \leq$$

$$\frac{\left[\|g(A)\| \left\|f^{(n)}\right\|_{\infty,[m,M]} + \|f(A)\| \left\|g^{(n)}\right\|_{\infty,[m,M]}\right]}{2(M-m)}$$

$$\|P_{n-1}\|_{\infty,[m,M]} \left[\left\|(M1_H - A)^2\right\| + \left\|(A - m1_H)^2\right\|\right].$$

Proof Here $\{E_\lambda\}_{\lambda \in \mathbb{R}}$ is the spectral family of A. Set

$$k(\lambda, t) := \begin{cases} t - m, & m \leq t \leq \lambda, \\ t - M, & \lambda < t \leq M. \end{cases} \tag{9.25}$$

where $\lambda \in [m, M]$.

Hence by (9.23) we obtain

$$f(\lambda) = \sum_{k=1}^{n-1} (-1)^{k+1} P_k(\lambda) f^{(k)}(\lambda) + \tag{9.26}$$

$$\sum_{k=1}^{n-1} \frac{(-1)^k (n-k)}{M-m} \left[P_k(M) f^{(k-1)}(M) - P_k(m) f^{(k-1)}(m)\right] +$$

$$\frac{n}{M-m} \int_m^M f(t)\,dt + \frac{(-1)^{n+1}}{M-m} \int_m^M P_{n-1}(t) k(\lambda, t) f^{(n)}(t)\,dt,$$

and

$$g(\lambda) = \sum_{k=1}^{n-1} (-1)^{k+1} P_k(\lambda) g^{(k)}(\lambda) + \tag{9.27}$$

$$\sum_{k=1}^{n-1}\frac{(-1)^k(n-k)}{M-m}\left[P_k(M)g^{(k-1)}(M)-P_k(m)g^{(k-1)}(m)\right]+$$

$$\frac{n}{M-m}\int_m^M g(t)\,dt+\frac{(-1)^{n+1}}{M-m}\int_m^M P_{n-1}(t)\,k(\lambda,t)\,g^{(n)}(t)\,dt,$$

$\forall\,\lambda\in[m,M]$.

By applying the spectral representation theorem on (9.26) and (9.27), i.e. integrating against E_λ over $[m,M]$, see (9.13), (ii), we obtain:

$$f(A)=\sum_{k=1}^{n-1}(-1)^{k+1}P_k(A)\,f^{(k)}(A)+ \tag{9.28}$$

$$\left(\sum_{k=1}^{n-1}\frac{(-1)^k(n-k)}{M-m}\left[P_k(M)f^{(k-1)}(M)-P_k(m)f^{(k-1)}(m)\right]\right)1_H+$$

$$\left(\frac{n}{M-m}\int_m^M f(t)\,dt\right)1_H+\frac{(-1)^{n+1}}{M-m}\int_{m-0}^M\left(\int_m^M P_{n-1}(t)\,k(\lambda,t)\,f^{(n)}(t)\,dt\right)dE_\lambda,$$

and

$$g(A)=\sum_{k=1}^{n-1}(-1)^{k+1}P_k(A)\,g^{(k)}(A)+ \tag{9.29}$$

$$\left(\sum_{k=1}^{n-1}\frac{(-1)^k(n-k)}{M-m}\left[P_k(M)g^{(k-1)}(M)-P_k(m)g^{(k-1)}(m)\right]\right)1_H+$$

$$\left(\frac{n}{M-m}\int_m^M g(t)\,dt\right)1_H+\frac{(-1)^{n+1}}{M-m}\int_{m-0}^M\left(\int_m^M P_{n-1}(t)\,k(\lambda,t)\,g^{(n)}(t)\,dt\right)dE_\lambda.$$

We notice that

$$g(A)\,f(A)=f(A)\,g(A) \tag{9.30}$$

to be used next.

Then it holds

$$g(A)\,f(A)=\sum_{k=1}^{n-1}(-1)^{k+1}g(A)\,P_k(A)\,f^{(k)}(A)+ \tag{9.31}$$

$$\left(\sum_{k=1}^{n-1}\frac{(-1)^k(n-k)}{M-m}\left[P_k(M)f^{(k-1)}(M)-P_k(m)f^{(k-1)}(m)\right]\right)g(A)+$$

$$\left(\frac{n}{M-m}\int_m^M f(t)\,dt\right) g(A) +$$

$$\frac{(-1)^{n+1}}{M-m} g(A) \int_{m-0}^M \left(\int_m^M P_{n-1}(t)\,k(\lambda,t)\,f^{(n)}(t)\,dt\right) dE_\lambda,$$

and

$$f(A)\,g(A) = \sum_{k=1}^{n-1} (-1)^{k+1} f(A)\,P_k(A)\,g^{(k)}(A) + \tag{9.32}$$

$$\left(\sum_{k=1}^{n-1} \frac{(-1)^k (n-k)}{M-m}\left[P_k(M)\,g^{(k-1)}(M) - P_k(m)\,g^{(k-1)}(m)\right]\right) f(A) +$$

$$\left(\frac{n}{M-m}\int_m^M g(t)\,dt\right) f(A) +$$

$$\frac{(-1)^{n+1}}{M-m} f(A) \int_{m-0}^M \left(\int_m^M P_{n-1}(t)\,k(\lambda,t)\,g^{(n)}(t)\,dt\right) dE_\lambda.$$

Here from now on we consider $x \in H : \|x\| = 1$; immediately we get $\int_{m-0}^M d\langle E_\lambda x, x\rangle = 1$.

Then it holds (see (9.12))

$$\langle f(A)\,x, x\rangle = \sum_{k=1}^{n-1} (-1)^{k+1} \left\langle P_k(A)\,f^{(k)}(A)\,x, x\right\rangle + \tag{9.33}$$

$$\sum_{k=1}^{n-1} \frac{(-1)^k (n-k)}{M-m}\left[P_k(M)\,f^{(k-1)}(M) - P_k(m)\,f^{(k-1)}(m)\right] +$$

$$\frac{n}{M-m}\int_m^M f(t)\,dt + \frac{(-1)^{n+1}}{M-m}\int_{m-0}^M \left(\int_m^M P_{n-1}(t)\,k(\lambda,t)\,f^{(n)}(t)\,dt\right) d\langle E_\lambda x, x\rangle,$$

and

$$\langle g(A)\,x, x\rangle = \sum_{k=1}^{n-1} (-1)^{k+1} \left\langle P_k(A)\,g^{(k)}(A)\,x, x\right\rangle + \tag{9.34}$$

$$\sum_{k=1}^{n-1} \frac{(-1)^k (n-k)}{M-m}\left[P_k(M)\,g^{(k-1)}(M) - P_k(m)\,g^{(k-1)}(m)\right] +$$

$$\frac{n}{M-m}\int_m^M g(t)\,dt + \frac{(-1)^{n+1}}{M-m}\int_{m-0}^M \left(\int_m^M P_{n-1}(t)\,k(\lambda,t)\,g^{(n)}(t)\,dt\right) d\langle E_\lambda x, x\rangle.$$

Then we get

$$\langle f(A)x,x\rangle\langle g(A)x,x\rangle=\sum_{k=1}^{n-1}(-1)^{k+1}\left\langle P_k(A)f^{(k)}(A)x,x\right\rangle\langle g(A)x,x\rangle+ \tag{9.35}$$
$$\left(\sum_{k=1}^{n-1}\frac{(-1)^k(n-k)}{M-m}\left[P_k(M)f^{(k-1)}(M)-P_k(m)f^{(k-1)}(m)\right]\right)\langle g(A)x,x\rangle+$$
$$\left(\frac{n}{M-m}\int_m^M f(t)\,dt\right)\langle g(A)x,x\rangle+$$
$$\frac{(-1)^{n+1}\langle g(A)x,x\rangle}{M-m}\int_{m-0}^{M}\left(\int_m^M P_{n-1}(t)k(\lambda,t)f^{(n)}(t)\,dt\right)d\langle E_\lambda x,x\rangle,$$

and

$$\langle g(A)x,x\rangle\langle f(A)x,x\rangle=\sum_{k=1}^{n-1}(-1)^{k+1}\left\langle P_k(A)g^{(k)}(A)x,x\right\rangle\langle f(A)x,x\rangle+ \tag{9.36}$$
$$\left(\sum_{k=1}^{n-1}\frac{(-1)^k(n-k)}{M-m}\left[P_k(M)g^{(k-1)}(M)-P_k(m)g^{(k-1)}(m)\right]\right)\langle f(A)x,x\rangle+$$
$$\left(\frac{n}{M-m}\int_m^M g(t)\,dt\right)\langle f(A)x,x\rangle+$$
$$\frac{(-1)^{n+1}\langle f(A)x,x\rangle}{M-m}\int_{m-0}^{M}\left(\int_m^M P_{n-1}(t)k(\lambda,t)g^{(n)}(t)\,dt\right)d\langle E_\lambda x,x\rangle.$$

Furthermore we obtain

$$\langle f(A)g(A)x,x\rangle\overset{(9.31)}{=}\sum_{k=1}^{n-1}(-1)^{k+1}\left\langle g(A)P_k(A)f^{(k)}(A)x,x\right\rangle+ \tag{9.37}$$
$$\left(\sum_{k=1}^{n-1}\frac{(-1)^k(n-k)}{M-m}\left[P_k(M)f^{(k-1)}(M)-P_k(m)f^{(k-1)}(m)\right]\right)\langle g(A)x,x\rangle+$$
$$\left(\frac{n}{M-m}\int_m^M f(t)\,dt\right)\langle g(A)x,x\rangle+$$
$$\frac{(-1)^{n+1}}{M-m}\left\langle\left(g(A)\int_{m-0}^{M}\left(\int_m^M P_{n-1}(t)k(\lambda,t)f^{(n)}(t)\,dt\right)dE_\lambda\right)x,x\right\rangle,$$

and

$$\langle f(A) g(A) x, x\rangle \overset{(9.32)}{=} \sum_{k=1}^{n-1} (-1)^{k+1} \left\langle f(A) P_k(A) g^{(k)}(A) x, x\right\rangle + \tag{9.38}$$

$$\left(\sum_{k=1}^{n-1} \frac{(-1)^k (n-k)}{M-m} \left[P_k(M) g^{(k-1)}(M) - P_k(m) g^{(k-1)}(m)\right]\right) \langle f(A) x, x\rangle +$$

$$\left(\frac{n}{M-m} \int_m^M g(t)\, dt\right) \langle f(A) x, x\rangle +$$

$$\frac{(-1)^{n+1}}{M-m} \left\langle \left(f(A) \int_{m-0}^M \left(\int_m^M P_{n-1}(t) k(\lambda, t) g^{(n)}(t)\, dt\right) dE_\lambda\right) x, x\right\rangle .$$

By (9.35) and (9.37) we obtain

$$E := \langle f(A) g(A) x, x\rangle - \langle f(A) x, x\rangle \langle g(A) x, x\rangle = \tag{9.39}$$

$$\sum_{k=1}^{n-1} (-1)^{k+1} \left[\left\langle g(A) P_k(A) f^{(k)}(A) x, x\right\rangle - \left\langle P_k(A) f^{(k)}(A) x, x\right\rangle \langle g(A) x, x\rangle\right] +$$

$$\frac{(-1)^{n+1}}{M-m} \left[\left\langle \left(g(A) \int_{m-0}^M \left(\int_m^M P_{n-1}(t) k(\lambda, t) f^{(n)}(t)\, dt\right) dE_\lambda\right) x, x\right\rangle - \right.$$

$$\left. \langle g(A) x, x\rangle \int_{m-0}^M \left(\int_m^M P_{n-1}(t) k(\lambda, t) f^{(n)}(t)\, dt\right) d\langle E_\lambda x, x\rangle\right],$$

and by (9.36) and (9.38) we derive

$$E := \langle f(A) g(A) x, x\rangle - \langle f(A) x, x\rangle \langle g(A) x, x\rangle = \tag{9.40}$$

$$\sum_{k=1}^{n-1} (-1)^{k+1} \left[\left\langle f(A) P_k(A) g^{(k)}(A) x, x\right\rangle - \left\langle P_k(A) g^{(k)}(A) x, x\right\rangle \langle f(A) x, x\rangle\right] +$$

$$\frac{(-1)^{n+1}}{M-m} \left[\left\langle \left(f(A) \int_{m-0}^M \left(\int_m^M P_{n-1}(t) k(\lambda, t) g^{(n)}(t)\, dt\right) dE_\lambda\right) x, x\right\rangle - \right.$$

$$\left. \langle f(A) x, x\rangle \int_{m-0}^M \left(\int_m^M P_{n-1}(t) k(\lambda, t) g^{(n)}(t)\, dt\right) d\langle E_\lambda x, x\rangle\right].$$

Consequently, we get that

$$2E=\sum_{k=1}^{n-1}(-1)^{k+1}\left\{\left[\left\langle g(A)P_k(A)f^{(k)}(A)x,x\right\rangle+\left\langle f(A)P_k(A)g^{(k)}(A)x,x\right\rangle\right]-\right.$$

$$\left.\left[\left\langle P_k(A)f^{(k)}(A)x,x\right\rangle\langle g(A)x,x\rangle+\left\langle P_k(A)g^{(k)}(A)x,x\right\rangle\langle f(A)x,x\rangle\right]\right\}+$$

$$\frac{(-1)^{n+1}}{M-m}\left\{\left[\left\langle\left(g(A)\int_{m-0}^{M}\left(\int_m^M P_{n-1}(t)k(\lambda,t)f^{(n)}(t)dt\right)dE_\lambda\right)x,x\right\rangle+\right.\right.$$

$$\left.\left\langle\left(f(A)\int_{m-0}^{M}\left(\int_m^M P_{n-1}(t)k(\lambda,t)g^{(n)}(t)dt\right)dE_\lambda\right)x,x\right\rangle\right]-$$

$$\left[\langle g(A)x,x\rangle\int_{m-0}^{M}\left(\int_m^M P_{n-1}(t)k(\lambda,t)f^{(n)}(t)dt\right)d\langle E_\lambda x,x\rangle+\right.$$

$$\left.\left.\langle f(A)x,x\rangle\int_{m-0}^{M}\left(\int_m^M P_{n-1}(t)k(\lambda,t)g^{(n)}(t)dt\right)d\langle E_\lambda x,x\rangle\right]\right\}. \quad (9.41)$$

We find that

$$\langle f(A)g(A)x,x\rangle-\langle f(A)x,x\rangle\langle g(A)x,x\rangle-$$

$$\frac{1}{2}\left[\sum_{k=1}^{n-1}(-1)^{k+1}\left\{\left[\left\langle P_k(A)\left(g(A)f^{(k)}(A)+f(A)g^{(k)}(A)\right)x,x\right\rangle\right]-\right.\right.$$

$$\left.\left.\left[\left\langle P_k(A)f^{(k)}(A)x,x\right\rangle\langle g(A)x,x\rangle+\left\langle P_k(A)g^{(k)}(A)x,x\right\rangle\langle f(A)x,x\rangle\right]\right\}\right]=$$

$$\frac{(-1)^{n+1}}{2(M-m)}\left\{\left[\left\langle\left(g(A)\int_{m-0}^{M}\left(\int_m^M P_{n-1}(t)k(\lambda,t)f^{(n)}(t)dt\right)dE_\lambda\right)x,x\right\rangle+\right.\right.$$

$$\left.\left\langle\left(f(A)\int_{m-0}^{M}\left(\int_m^M P_{n-1}(t)k(\lambda,t)g^{(n)}(t)dt\right)dE_\lambda\right)x,x\right\rangle\right]-$$

$$\left[\langle g(A)x,x\rangle\int_{m-0}^{M}\left(\int_m^M P_{n-1}(t)k(\lambda,t)f^{(n)}(t)dt\right)d\langle E_\lambda x,x\rangle+\right.$$

$$\left.\left.\langle f(A)x,x\rangle\int_{m-0}^{M}\left(\int_m^M P_{n-1}(t)k(\lambda,t)g^{(n)}(t)dt\right)d\langle E_\lambda x,x\rangle\right]\right\}=:R. \quad (9.42)$$

Therefore it holds

$$|R| \leq \frac{1}{2(M-m)} \left\{ \left[\|g(A)\| \left\| \int_{m-0}^{M} \left(\int_{m}^{M} P_{n-1}(t) k(\lambda, t) f^{(n)}(t) dt \right) dE_{\lambda} \right\| + \right. \right.$$

$$\left. \|f(A)\| \left\| \int_{m-0}^{M} \left(\int_{m}^{M} P_{n-1}(t) k(\lambda, t) g^{(n)}(t) dt \right) dE_{\lambda} \right\| \right] + \tag{9.43}$$

$$\left[\|g(A)\| \left\| \int_{m-0}^{M} \left(\int_{m}^{M} P_{n-1}(t) k(\lambda, t) f^{(n)}(t) dt \right) dE_{\lambda} \right\| + \right.$$

$$\left. \left. \|f(A)\| \left\| \int_{m-0}^{M} \left(\int_{m}^{M} P_{n-1}(t) k(\lambda, t) g^{(n)}(t) dt \right) dE_{\lambda} \right\| \right] \right\} =$$

$$\frac{1}{(M-m)} \left\{ \|g(A)\| \left\| \int_{m-0}^{M} \left(\int_{m}^{M} P_{n-1}(t) k(\lambda, t) f^{(n)}(t) dt \right) dE_{\lambda} \right\| + \right.$$

$$\left. \|f(A)\| \left\| \int_{m-0}^{M} \left(\int_{m}^{M} P_{n-1}(t) k(\lambda, t) g^{(n)}(t) dt \right) dE_{\lambda} \right\| \right\} =: (\xi_1). \tag{9.44}$$

We notice the following:

$$\left\| \int_{m-0}^{M} \left(\int_{m}^{M} P_{n-1}(t) k(\lambda, t) f^{(n)}(t) dt \right) dE_{\lambda} \right\| =$$

$$\sup_{x \in H: \|x\|=1} \left| \int_{m-0}^{M} \left(\int_{m}^{M} P_{n-1}(t) k(\lambda, t) f^{(n)}(t) dt \right) d\langle E_{\lambda} x, x \rangle \right| \leq$$

$$\sup_{x \in H: \|x\|=1} \left(\int_{m-0}^{M} \left(\int_{m}^{M} |P_{n-1}(t)| |k(\lambda, t)| \left| f^{(n)}(t) \right| dt \right) d\langle E_{\lambda} x, x \rangle \right) \leq \tag{9.45}$$

$$\left(\|P_{n-1}\|_{\infty,[m,M]} \left\| f^{(n)} \right\|_{\infty,[m,M]} \right)$$

$$\sup_{x \in H: \|x\|=1} \left(\int_{m-0}^{M} \left(\int_{m}^{M} |k(\lambda, t)| dt \right) d\langle E_{\lambda} x, x \rangle \right) =: (\xi_2).$$

Notice that

$$\int_{m}^{M} |k(\lambda, t)| dt = \int_{m}^{\lambda} (t-m) dt + \int_{\lambda}^{M} (M-t) dt = \frac{(\lambda-m)^2 + (M-\lambda)^2}{2}. \tag{9.46}$$

Hence it holds

$$(\xi_2) \overset{(9.46)}{=} \left(\frac{\|P_{n-1}\|_{\infty,[m,M]} \left\| f^{(n)} \right\|_{\infty,[m,M]}}{2} \right).$$

$$\sup_{x \in H: \|x\|=1} \left[\left\langle (M1_H - A)^2 x, x \right\rangle + \left\langle (A - m1_H)^2 x, x \right\rangle \right] \leq$$

$$\left(\frac{\|P_{n-1}\|_{\infty,[m,M]} \left\| f^{(n)} \right\|_{\infty,[m,M]}}{2} \right) \left[\left\| (M1_H - A)^2 \right\| + \left\| (A - m1_H)^2 \right\| \right]. \quad (9.47)$$

We have proved that

$$\left\| \int_{m-0}^{M} \left(\int_{m}^{M} P_{n-1}(t)\, k(\lambda, t)\, f^{(n)}(t)\, dt \right) dE_\lambda \right\| \leq \quad (9.48)$$

$$\left(\frac{\|P_{n-1}\|_{\infty,[m,M]} \left\| f^{(n)} \right\|_{\infty,[m,M]}}{2} \right) \left[\left\| (M1_H - A)^2 \right\| + \left\| (A - m1_H)^2 \right\| \right].$$

Similarly, it holds

$$\left\| \int_{m-0}^{M} \left(\int_{m}^{M} P_{n-1}(t)\, k(\lambda, t)\, g^{(n)}(t)\, dt \right) dE_\lambda \right\| \leq$$

$$\left(\frac{\|P_{n-1}\|_{\infty,[m,M]} \left\| g^{(n)} \right\|_{\infty,[m,M]}}{2} \right) \left[\left\| (M1_H - A)^2 \right\| + \left\| (A - m1_H)^2 \right\| \right]. \quad (9.49)$$

Next we apply (9.48) and (9.49) into (9.44), we get

$$(\xi_1) \leq \frac{1}{(M-m)} \left\{ \|g(A)\| \left(\frac{\|P_{n-1}\|_{\infty,[m,M]} \left\| f^{(n)} \right\|_{\infty,[m,M]}}{2} \cdot \right. \right. \quad (9.50)$$

$$\left. \left[\left\| (M1_H - A)^2 \right\| + \left\| (A - m1_H)^2 \right\| \right] \right) + \|f(A)\| \cdot$$

$$\left. \left(\frac{\|P_{n-1}\|_{\infty,[m,M]} \left\| g^{(n)} \right\|_{\infty,[m,M]}}{2} \right) \left[\left\| (M1_H - A)^2 \right\| + \left\| (A - m1_H)^2 \right\| \right] \right\} =$$

$$\frac{1}{2(M-m)} \left\{ \left[\|g(A)\| \left\| f^{(n)} \right\|_{\infty,[m,M]} + \|f(A)\| \left\| g^{(n)} \right\|_{\infty,[m,M]} \right] \right.$$

$$\left. \|P_{n-1}\|_{\infty,[m,M]} \left[\left\| (M1_H - A)^2 \right\| + \left\| (A - m1_H)^2 \right\| \right] \right\}. \quad (9.51)$$

We have proved that

$$|R| \leq \frac{\left(\|g(A)\| \left\| f^{(n)} \right\|_{\infty,[m,M]} + \|f(A)\| \left\| g^{(n)} \right\|_{\infty,[m,M]} \right)}{2(M-m)}$$

$$\|P_{n-1}\|_{\infty,[m,M]} \left[\left\| (M1_H - A)^2 \right\| + \left\| (A - m1_H)^2 \right\| \right]. \tag{9.52}$$

The theorem is proved. ■

It follows the case $n = 1$.

Corollary 9.7 (to Theorem 9.6). *Let $f, g \in C^1([a, b])$ with $[m, M] \subset (a, b)$, $m < M$. Here A is a selfadjoint bounded linear operator on the Hilbert space H with spectrum $Sp(A) \subseteq [m, M]$. We consider any $x \in H : \|x\| = 1$.*

Then

$$|\langle f(A) g(A) x, x \rangle - \langle f(A) x, x \rangle \langle g(A) x, x \rangle| \leq \tag{9.53}$$

$$\frac{\left[\|g(A)\| \left\| f' \right\|_{\infty,[m,M]} + \|f(A)\| \left\| g' \right\|_{\infty,[m,M]} \right]}{2(M-m)}$$

$$\left[\left\| (M1_H - A)^2 \right\| + \left\| (A - m1_H)^2 \right\| \right].$$

We continue with

Theorem 9.8 *All as in Theorem 9.6. Let $\alpha, \beta, \gamma > 1 : \frac{1}{\alpha} + \frac{1}{\beta} + \frac{1}{\gamma} = 1$. Then*

$$\langle (\Delta(f, g))(A) x, x \rangle \leq \frac{\|P_{n-1}\|_{\alpha,[m,M]}}{(M-m)(\beta+1)^{\frac{1}{\beta}}}$$

$$\left[\|g(A)\| \left\| f^{(n)} \right\|_{\gamma,[m,M]} + \|f(A)\| \left\| g^{(n)} \right\|_{\gamma,[m,M]} \right] \tag{9.54}$$

$$\left[\left\| (A - m1_H)^{1+\frac{1}{\beta}} \right\| + \left\| (M1_H - A)^{1+\frac{1}{\beta}} \right\| \right].$$

Proof As in (9.45) we have

$$\left\| \int_{m-0}^{M} \left(\int_{m}^{M} P_{n-1}(t) k(\lambda, t) f^{(n)}(t) dt \right) dE_\lambda \right\| =$$

$$\sup_{x \in H : \|x\| = 1} \left| \int_{m-0}^{M} \left(\int_{m}^{M} P_{n-1}(t) k(\lambda, t) f^{(n)}(t) dt \right) d(\langle E_\lambda x, x \rangle) \right| =: \psi_1. \tag{9.55}$$

Here $\alpha, \beta, \gamma > 1 : \frac{1}{\alpha} + \frac{1}{\beta} + \frac{1}{\gamma} = 1$. By Hölder's inequality for three functions we get

$$\left| \int_m^M P_{n-1}(t) k(\lambda, t) f^{(n)}(t) dt \right| \leq \int_m^M |P_{n-1}(t)| |k(\lambda, t)| \left| f^{(n)}(t) \right| dt \leq$$

$$\|P_{n-1}\|_\alpha \left\| f^{(n)} \right\|_\gamma \left(\int_m^M |k(\lambda, t)|^\beta dt \right)^{\frac{1}{\beta}} =$$

$$\|P_{n-1}\|_\alpha \left\| f^{(n)} \right\|_\gamma \left(\int_m^\lambda (t-m)^\beta dt + \int_\lambda^M (M-t)^\beta dt \right)^{\frac{1}{\beta}} = \tag{9.56}$$

$$\|P_{n-1}\|_\alpha \left\| f^{(n)} \right\|_\gamma \left[\frac{(\lambda - m)^{\beta+1} + (M - \lambda)^{\beta+1}}{\beta + 1} \right]^{\frac{1}{\beta}} \leq$$

$$\frac{\|P_{n-1}\|_\alpha \left\| f^{(n)} \right\|_\gamma}{(\beta+1)^{\frac{1}{\beta}}} \left[(\lambda - m)^{\frac{\beta+1}{\beta}} + (M - \lambda)^{\frac{\beta+1}{\beta}} \right].$$

I.e. it holds

$$\left| \int_m^M P_{n-1}(t) k(\lambda, t) f^{(n)}(t) dt \right| \leq$$

$$\frac{\|P_{n-1}\|_\alpha \left\| f^{(n)} \right\|_\gamma}{(\beta+1)^{\frac{1}{\beta}}} \left[(\lambda - m)^{1+\frac{1}{\beta}} + (M - \lambda)^{1+\frac{1}{\beta}} \right], \quad \forall \lambda \in [m, M]. \tag{9.57}$$

Therefore we get

$$\psi_1 \leq \sup_{x \in H: \|x\|=1} \int_{m-0}^M \left| \int_m^M P_{n-1}(t) k(\lambda, t) f^{(n)}(t) dt \right| d\langle E_\lambda x, x \rangle \leq$$

$$\left(\sup_{x \in H: \|x\|=1} \int_{m-0}^M \left[(\lambda - m)^{1+\frac{1}{\beta}} + (M - \lambda)^{1+\frac{1}{\beta}} \right] d\langle E_\lambda x, x \rangle \right)$$

$$\frac{\|P_{n-1}\|_{\alpha,[m,M]} \left\| f^{(n)} \right\|_{\gamma,[m,M]}}{(\beta+1)^{\frac{1}{\beta}}} \leq$$

$$\left(\frac{\|P_{n-1}\|_{\alpha,[m,M]} \left\| f^{(n)} \right\|_{\gamma,[m,M]}}{(\beta+1)^{\frac{1}{\beta}}} \right) \tag{9.58}$$

$$\left[\left\|(A-m1_H)^{1+\frac{1}{\beta}}\right\|+\left\|(M1_H-A)^{1+\frac{1}{\beta}}\right\|\right].$$

We have proved that

$$\left\|\int_{m-0}^{M}\left(\int_{m}^{M}P_{n-1}(t)\,k(\lambda,t)\,f^{(n)}(t)\,dt\right)dE_{\lambda}\right\|\leq \tag{9.59}$$

$$\frac{\|P_{n-1}\|_{\alpha,[m,M]}\left\|f^{(n)}\right\|_{\gamma,[m,M]}}{(\beta+1)^{\frac{1}{\beta}}}\left[\left\|(A-m1_H)^{1+\frac{1}{\beta}}\right\|+\left\|(M1_H-A)^{1+\frac{1}{\beta}}\right\|\right].$$

Similarly, it holds

$$\left\|\int_{m-0}^{M}\left(\int_{m}^{M}P_{n-1}(t)\,k(\lambda,t)\,g^{(n)}(t)\,dt\right)dE_{\lambda}\right\|\leq$$

$$\frac{\|P_{n-1}\|_{\alpha,[m,M]}\left\|g^{(n)}\right\|_{\gamma,[m,M]}}{(\beta+1)^{\frac{1}{\beta}}}\left[\left\|(A-m1_H)^{1+\frac{1}{\beta}}\right\|+\left\|(M1_H-A)^{1+\frac{1}{\beta}}\right\|\right]. \tag{9.60}$$

Using (9.44) we derive

$$|R|\leq\frac{1}{(M-m)}\left\{\|g(A)\|\frac{\|P_{n-1}\|_{\alpha,[m,M]}\left\|f^{(n)}\right\|_{\gamma,[m,M]}}{(\beta+1)^{\frac{1}{\beta}}}\right. \tag{9.61}$$

$$\left[\left\|(A-m1_H)^{1+\frac{1}{\beta}}\right\|+\left\|(M1_H-A)^{1+\frac{1}{\beta}}\right\|\right]+$$

$$\|f(A)\|\frac{\|P_{n-1}\|_{\alpha,[m,M]}\left\|g^{(n)}\right\|_{\gamma,[m,M]}}{(\beta+1)^{\frac{1}{\beta}}}$$

$$\left.\left[\left\|(A-m1_H)^{1+\frac{1}{\beta}}\right\|+\left\|(M1_H-A)^{1+\frac{1}{\beta}}\right\|\right]\right\}=$$

$$\frac{1}{(M-m)}\left[\|g(A)\|\left\|f^{(n)}\right\|_{\gamma,[m,M]}+\|f(A)\|\left\|g^{(n)}\right\|_{\gamma,[m,M]}\right]\frac{\|P_{n-1}\|_{\alpha,[m,M]}}{(\beta+1)^{\frac{1}{\beta}}} \tag{9.62}$$

$$\left[\left\|(A-m1_H)^{1+\frac{1}{\beta}}\right\|+\left\|(M1_H-A)^{1+\frac{1}{\beta}}\right\|\right],$$

proving the claim. ■

The case $n = 1$ follows.

Corollary 9.9 (to Theorem 9.8). *All as in Theorem 9.8. It holds*

$$|\langle f(A) g(A) x, x\rangle - \langle f(A) x, x\rangle \langle g(A) x, x\rangle| \leq$$

$$\frac{1}{(M-m)(\beta+1)^{\frac{1}{\beta}}} \left[\|g(A)\| \left\|f'\right\|_{\gamma,[m,M]} + \|f(A)\| \left\|g'\right\|_{\gamma,[m,M]}\right] \tag{9.63}$$

$$\left[\left\|(A - m1_H)^{1+\frac{1}{\beta}}\right\| + \left\|(M1_H - A)^{1+\frac{1}{\beta}}\right\|\right].$$

We also give

Theorem 9.10 *All as in Theorem 9.6. It holds*

$$\langle(\Delta(f,g))(A) x, x\rangle \leq \|P_{n-1}\|_{\infty,[m,M]}$$

$$\left[\|g(A)\| \left\|f^{(n)}\right\|_{1,[m,M]} + \|f(A)\| \left\|g^{(n)}\right\|_{1,[m,M]}\right]. \tag{9.64}$$

Proof We have that

$$\left|\int_m^M P_{n-1}(t) k(\lambda,t) f^{(n)}(t) dt\right| \leq \int_m^M |P_{n-1}(t)| |k(\lambda,t)| \left|f^{(n)}(t)\right| dt \leq$$

$$\|P_{n-1}\|_{\infty,[m,M]} (M-m) \int_m^M \left|f^{(n)}(t)\right| dt =$$

$$\|P_{n-1}\|_{\infty,[m,M]} (M-m) \left\|f^{(n)}\right\|_{1,[m,M]}. \tag{9.65}$$

So that

$$\left|\int_m^M P_{n-1}(t) k(\lambda,t) f^{(n)}(t) dt\right| \leq$$

$$(M-m) \|P_{n-1}\|_{\infty,[m,M]} \left\|f^{(n)}\right\|_{1,[m,M]}.$$

Hence

$$\left\|\int_{m-0}^M \left(\int_m^M P_{n-1}(t) k(\lambda,t) f^{(n)}(t) dt\right) dE_\lambda\right\| =$$

$$\sup_{x\in H:\|x\|=1} \left|\int_{m-0}^M \left(\int_m^M P_{n-1}(t) k(\lambda,t) f^{(n)}(t) dt\right) d\langle E_\lambda x, x\rangle\right| \leq \tag{9.66}$$

$$(M-m) \|P_{n-1}\|_{\infty,[m,M]} \left\|f^{(n)}\right\|_{1,[m,M]},$$

and similarly,

$$\left\| \int_{m-0}^{M} \left(\int_{m}^{M} P_{n-1}(t) k(\lambda, t) g^{(n)}(t) dt \right) dE_\lambda \right\| \leq \tag{9.67}$$

$$(M - m) \|P_{n-1}\|_{\infty,[m,M]} \left\| g^{(n)} \right\|_{1,[m,M]} .$$

Using (9.44) we obtain

$$|R| \leq \frac{1}{(M-m)} \left\{ \|g(A)\| (M-m) \|P_{n-1}\|_{\infty,[m,M]} \left\| f^{(n)} \right\|_{1,[m,M]} + \right.$$

$$\left. \|f(A)\| (M-m) \|P_{n-1}\|_{\infty,[m,M]} \left\| g^{(n)} \right\|_{1,[m,M]} \right\} =$$

$$\|P_{n-1}\|_{\infty,[m,M]} \left[\|g(A)\| \left\| f^{(n)} \right\|_{1,[m,M]} + \|f(A)\| \left\| g^{(n)} \right\|_{1,[m,M]} \right], \tag{9.68}$$

proving the claim. ■

The case $n = 1$ follows.

Corollary 9.11 (to Theorem 9.10). *It holds*

$$|\langle f(A) g(A) x, x \rangle - \langle f(A) x, x \rangle \langle g(A) x, x \rangle| \leq$$

$$\left[\|g(A)\| \left\| f' \right\|_{1,[m,M]} + \|f(A)\| \left\| g' \right\|_{1,[m,M]} \right]. \tag{9.69}$$

Comment 9.12 *The case of harmonic sequence of polynomials $P_k(t) = \frac{(t-x)^k}{k!}$, $k \in \mathbb{Z}_+$, was completely studied in [2], and this work generalizes it.*

Another harmonic sequence of polynomials related to this work is

$$P_k(t) = \frac{1}{k!} \left(t - \frac{m+M}{2} \right)^k, \quad k \in \mathbb{Z}_+, \tag{9.70}$$

see also [5].

The Bernoulli polynomials $B_n(t)$ can be defined by the formula (see [5])

$$\frac{x e^{tx}}{e^x - 1} = \sum_{n=0}^{\infty} \frac{B_n(t)}{n!} x^n, \quad |x| < 2\pi, \ t \in \mathbb{R}. \tag{9.71}$$

They satisfy the relation

$$B_n'(t) = n B_{n-1}(t), \quad n \in \mathbb{N}.$$

The sequence

$$P_n(t) = \frac{1}{n!} B_n(t), \quad n \in \mathbb{Z}_+, \tag{9.72}$$

is a harmonic sequence of polynomials, $t \in \mathbb{R}$.

The Euler polynomials are defined by the formula (see [5])

$$\frac{2e^{tx}}{e^x + 1} = \sum_{n=0}^{\infty} \frac{E_n(t)}{n!} x^n, \quad |x| < \pi, \ t \in \mathbb{R}. \tag{9.73}$$

They satisfy

$$E_n'(t) = n E_{n-1}(t), \quad n \in \mathbb{N}.$$

The sequence

$$P_n(t) = \frac{1}{n!} E_n(t), \quad n \in \mathbb{Z}_+, \ t \in \mathbb{R}, \tag{9.74}$$

is a harmonic sequence of polynomials.

Finally:

Comment 9.13 *One can apply (9.24), (9.54) and (9.64), for the harmonic sequences of polynomials defined by (9.70), (9.72) and (9.74).*

In particular, when (see (9.70))

$$P_n(t) = \frac{1}{n!} \left(t - \frac{m+M}{2} \right)^n, \quad n \in \mathbb{Z}_+, \tag{9.75}$$

we get

$$\| P_{n-1} \|_{\infty,[m,M]} = \frac{1}{(n-1)!} \left(\frac{M-m}{2} \right)^{n-1}, \tag{9.76}$$

and

$$\| P_{n-1} \|_{\alpha,[m,M]} = \frac{1}{(n-1)! \left(\alpha (n-1) + 1 \right)^{\frac{1}{\alpha}}} \left(\frac{(M-m)^{\alpha(n-1)+1}}{2^{\alpha(n-1)}} \right), \tag{9.77}$$

where $\alpha, \beta, \gamma > 1 : \frac{1}{\alpha} + \frac{1}{\beta} + \frac{1}{\gamma} = 1$.

References

1. G.A. Anastassiou, *Intelligent Comparisons: Analytic Inequalities* (Springer, New York, Heidelberg, 2016)
2. G.A. Anastassiou, *Self Adjoint Operator Chebyshev-Grüss Type Inequalities* (2016, submitted)

3. G. Anastassiou, *Self Adjoint Operator Harmonic Chebyshev-Grüss Inequalities* (2016, submitted)
4. P.L. Čebyšev, Sur les expressions approximatives des intégrales définies par les autres proses entre les mêmes limites. Proc. Math. Soc. Charkov **2**, 93–98 (1882)
5. L. Dedić, J.E. Pečarić, N. Ujević, On generalizations of Ostrowski inequality and some related results. Czechoslovak Math. J. **53**(1), 173–189 (2003)
6. S.S. Dragomir, Inequalities for functions of selfadjoint operators on Hilbert spaces (2011). ajmaa.org/RGMIA/monographs/InFuncOp.pdf
7. S. Dragomir, *Operator Inequalities of Ostrowski and Trapezoidal Type* (Springer, New York, 2012)
8. A.M. Fink, Bounds on the deviation of a function from its averages. Czechoslovak Math. J. **42**(117), 289–310 (1992)
9. T. Furuta, J. Mićić Hot, J. Pečaric, Y. Seo, Mond-Pečaric Method in Operator Inequalities. Inequalities for Bounded Selfadjoint Operators on a Hilbert Space. (Element, Zagreb, 2005)
10. G. Grüss, Über das Maximum des absoluten Betrages von $\left[\left(\frac{1}{b-a}\right)\int_a^b f(x)\,g(x)\,dx - \left(\frac{1}{(b-a)^2}\int_a^b f(x)\,dx\int_a^b g(x)\,dx\right)\right]$. Math. Z. **39**, 215–226 (1935)
11. G. Helmberg, *Introduction to Spectral Theory in Hilbert Space* (Wiley, New York, 1969)

Chapter 10
Ultra General Self Adjoint Operator Chebyshev-Grüss Type Inequalities

We demonstrate here most general self adjoint operator Chebyshev-Grüss type inequalities to all cases. We finish with applications. It follows [2].

10.1 Motivation

Here we mention the following interesting and motivating results.

Theorem 10.1 (Čebyšev 1882 [3]). *Let $f, g : [a, b] \to \mathbb{R}$ absolutely continuous functions. If $f', g' \in L_{\infty}([a, b])$, then*

$$\left|\frac{1}{b-a}\int_a^b f(x)\,g(x)\,dx - \left(\frac{1}{b-a}\int_a^b f(x)\,dx\right)\left(\frac{1}{b-a}\int_a^b g(x)\,dx\right)\right| \tag{10.1}$$

$$\leq \frac{1}{12}(b-a)^2 \left\|f'\right\|_{\infty} \left\|g'\right\|_{\infty}.$$

Also we mention

Theorem 10.2 (Grüss 1935 [7]). *Let f, g integrable functions from $[a, b]$ into $\mathbb{R}$, such that $m \leq f(x) \leq M$, $\rho \leq g(x) \leq \sigma$, for all $x \in [a, b]$, where $m, M, \rho, \sigma \in \mathbb{R}$. Then*

$$\left|\frac{1}{b-a}\int_a^b f(x)\,g(x)\,dx - \left(\frac{1}{b-a}\int_a^b f(x)\,dx\right)\left(\frac{1}{b-a}\int_a^b g(x)\,dx\right)\right| \tag{10.2}$$

$$\leq \frac{1}{4}(M-m)(\sigma-\rho).$$

G.A. Anastassiou, *Intelligent Comparisons II: Operator Inequalities and Approximations*, Studies in Computational Intelligence 699,
DOI 10.1007/978-3-319-51475-8_10

10.2 Background

Let A be a selfadjoint linear operator on a complex Hilbert space $(H; \langle \cdot, \cdot \rangle)$. The Gelfand map establishes a $*-$isometrically isomorphism Φ between the set $C(Sp(A))$ of all continuous functions defined on the spectrum of A, denoted $Sp(A)$, and the C^*-algebra $C^*(A)$ generated by A and the identity operator 1_H on H as follows (see e.g. [6, p. 3]):

For any $f, g \in C(Sp(A))$ and any $\alpha, \beta \in \mathbb{C}$ we have

(i) $\Phi(\alpha f + \beta g) = \alpha \Phi(f) + \beta \Phi(g)$;
(ii) $\Phi(fg) = \Phi(f)\Phi(g)$ (the operation composition is on the right) and $\Phi(\overline{f}) = (\Phi(f))^*$;
(iii) $\|\Phi(f)\| = \|f\| := \sup_{t \in Sp(A)} |f(t)|$;
(iv) $\Phi(f_0) = 1_H$ and $\Phi(f_1) = A$, where $f_0(t) = 1$ and $f_1(t) = t$, for $t \in Sp(A)$.

With this notation we define

$$f(A) := \Phi(f), \text{ for all } f \in C(Sp(A)),$$

and we call it the continuous functional calculus for a selfadjoint operator A.

If A is a selfadjoint operator and f is a real valued continuous function on $Sp(A)$ then $f(t) \geq 0$ for any $t \in Sp(A)$ implies that $f(A) \geq 0$, i.e. $f(A)$ is a positive operator on H. Moreover, if both f and g are real valued functions on $Sp(A)$ then the following important property holds:

(P) $f(t) \geq g(t)$ for any $t \in Sp(A)$, implies that $f(A) \geq g(A)$ in the operator order of $B(H)$.

Equivalently, we use (see [5], pp. 7–8):

Let U be a selfadjoint operator on the complex Hilbert space $(H, \langle \cdot, \cdot \rangle)$ with the spectrum $Sp(U)$ included in the interval $[m, M]$ for some real numbers $m < M$ and $\{E_\lambda\}_\lambda$ be its spectral family.

Then for any continuous function $f : [m, M] \to \mathbb{C}$, it is well known that we have the following spectral representation in terms of the Riemann-Stieljes integral:

$$\langle f(U)x, y \rangle = \int_{m-0}^{M} f(\lambda) \, d(\langle E_\lambda x, y \rangle), \tag{10.3}$$

for any $x, y \in H$. The function $g_{x,y}(\lambda) := \langle E_\lambda x, y \rangle$ is of bounded variation on the interval $[m, M]$, and

$$g_{x,y}(m - 0) = 0 \text{ and } g_{x,y}(M) = \langle x, y \rangle,$$

for any $x, y \in H$. Furthermore, it is known that $g_x(\lambda) := \langle E_\lambda x, x \rangle$ is increasing and right continuous on $[m, M]$.

In this chapter we will be using a lot the formula

$$\langle f(U)x, x\rangle = \int_{m-0}^{M} f(\lambda)\, d(\langle E_\lambda x, x\rangle), \ \forall x \in H. \tag{10.4}$$

As a symbol we can write

$$f(U) = \int_{m-0}^{M} f(\lambda)\, dE_\lambda. \tag{10.5}$$

Above, $m = \min\{\lambda|\lambda \in Sp(U)\} := \min Sp(U)$, $M = \max\{\lambda|\lambda \in Sp(U)\} := \max Sp(U)$. The projections $\{E_\lambda\}_{\lambda\in\mathbb{R}}$, are called the spectral family of A, with the properties:

(a) $E_\lambda \leq E_{\lambda'}$ for $\lambda \leq \lambda'$;

(b) $E_{m-0} = 0_H$ (zero operator), $E_M = 1_H$ (identity operator) and $E_{\lambda+0} = E_\lambda$ for all $\lambda \in \mathbb{R}$.

Furthermore

$$E_\lambda := \varphi_\lambda(U), \ \forall \lambda \in \mathbb{R}, \tag{10.6}$$

is a projection which reduces U, with

$$\varphi_\lambda(s) := \begin{cases} 1, & \text{for } -\infty < s \leq \lambda, \\ 0, & \text{for } \lambda < s < +\infty. \end{cases}$$

The spectral family $\{E_\lambda\}_{\lambda\in\mathbb{R}}$ determines uniquely the self-adjoint operator U and vice versa.

For more on the topic see [8], pp. 256–266, and for more details see there pp. 157–266, see also [4].

Some more basics are given (we follow [5], pp. 1–5):

Let $(H; \langle\cdot,\cdot\rangle)$ be a Hilbert space over $\mathbb{C}$. A bounded linear operator A defined on H is selfjoint, i.e., $A = A^*$, iff $\langle Ax, x\rangle \in \mathbb{R}$, $\forall x \in H$, and if A is selfadjoint, then

$$\|A\| = \sup_{x\in H:\|x\|=1} |\langle Ax, x\rangle|. \tag{10.7}$$

Let A, B be selfadjoint operators on H. Then $A \leq B$ iff $\langle Ax, x\rangle \leq \langle Bx, x\rangle, \forall x \in H$.

In particular, A is called positive if $A \geq 0$.

Denote by

$$\mathcal{P} := \left\{\varphi(s) := \sum_{k=0}^{n} \alpha_k s^k | n \geq 0, \alpha_k \in \mathbb{C}, 0 \leq k \leq n\right\}. \tag{10.8}$$

If $A \in \mathcal{B}(H)$ (the Banach algebra of all bounded linear operators defined on H, i.e. from H into itself) is selfadjoint, and $\varphi(s) \in \mathcal{P}$ has real coefficients, then $\varphi(A)$ is selfadjoint, and

$$\|\varphi(A)\| = \max\{|\varphi(\lambda)|, \lambda \in Sp(A)\}. \tag{10.9}$$

If φ is any function defined on $\mathbb{R}$ we define

$$\|\varphi\|_A := \sup\{|\varphi(\lambda)|, \lambda \in Sp(A)\}. \tag{10.10}$$

If A is selfadjoint operator on Hilbert space H and φ is continuous and given that $\varphi(A)$ is selfadjoint, then $\|\varphi(A)\| = \|\varphi\|_A$. And if φ is a continuous real valued function so it is $|\varphi|$, then $\varphi(A)$ and $|\varphi|(A) = |\varphi(A)|$ are selfadjoint operators (by [5], p. 4, Theorem 7).

Hence it holds

$$\||\varphi(A)|\| = \||\varphi|\|_A = \sup\{||\varphi(\lambda)||, \lambda \in Sp(A)\}$$

$$= \sup\{|\varphi(\lambda)|, \lambda \in Sp(A)\} = \|\varphi\|_A = \|\varphi(A)\|,$$

that is

$$\||\varphi(A)|\| = \|\varphi(A)\|. \tag{10.11}$$

For a selfadjoint operator $A \in \mathcal{B}(H)$ which is positive, there exists a unique positive selfadjoint operator $B := \sqrt{A} \in \mathcal{B}(H)$ such that $B^2 = A$, that is $\left(\sqrt{A}\right)^2 = A$. We call B the square root of A.

Let $A \in \mathcal{B}(H)$, then A^*A is selfadjoint and positive. Define the "operator absolute value" $|A| := \sqrt{A^*A}$. If $A = A^*$, then $|A| = \sqrt{A^2}$.

For a continuous real valued function φ we observe the following:

$$|\varphi(A)| \text{ (the functional absolute value) } = \int_{m-0}^{M} |\varphi(\lambda)| \, dE_\lambda =$$

$$\int_{m-0}^{M} \sqrt{(\varphi(\lambda))^2} dE_\lambda = \sqrt{(\varphi(A))^2} = |\varphi(A)| \text{ (operator absolute value)},$$

where A is a selfadjoint operator.

That is we have

$$|\varphi(A)| \text{ (functional absolute value) } = |\varphi(A)| \text{ (operator absolute value)}. \tag{10.12}$$

Let $A, B \in \mathcal{B}(H)$, then

$$\|AB\| \leq \|A\| \, \|B\|, \tag{10.13}$$

by Banach algebra property.

10.3 Main Results

Next we present most general Chebyshev-Grüss type operator inequalities based on Theorem 26.9 of [1], p. 404.

Then we specialize them for $n = 1$.

We give

Theorem 10.3 *Let $n \in \mathbb{N}$ and $f_1, f_2 \in C^n([a,b])$ with $[m, M] \subset (a,b)$, $m < M$; $g \in C^1([a,b])$ and $g^{-1} \in C^n([a,b])$. Here A is a selfadjoint linear operator on the Hilbert space H with spectrum $Sp(A) \subseteq [m, M]$. We consider any $x \in H : \|x\| = 1$. Then*

$$\langle (\Delta(f_1, f_2; g))(A)x, x\rangle :=$$

$$\Bigg| \langle f_1(A) f_2(A)x, x\rangle - \langle f_1(A)x, x\rangle \cdot \langle f_2(A)x, x\rangle - \frac{1}{2(M-m)} \left\{ \sum_{k=1}^{n-1} \frac{1}{k!} \cdot \right.$$

$$\left\{ \left[\left\langle \left(f_2(A) \int_{m-0}^{M} \left(\int_m^M \left(f_1 \circ g^{-1}\right)^{(k)} (g(t)) (g(\lambda) - g(t))^k dt \right) dE_\lambda \right) x, x \right\rangle - \right. \right.$$

$$\left. \langle f_2(A)x, x\rangle \int_{m-0}^{M} \left(\int_m^M \left(f_1 \circ g^{-1}\right)^{(k)} (g(t)) (g(\lambda) - g(t))^k dt \right) d\langle E_\lambda x, x\rangle \right] +$$

$$\left[\left\langle \left(f_1(A) \int_{m-0}^{M} \left(\int_m^M \left(f_2 \circ g^{-1}\right)^{(k)} (g(t)) (g(\lambda) - g(t))^k dt \right) dE_\lambda \right) x, x \right\rangle - \right.$$

$$\left. \left. \left. \langle f_1(A)x, x\rangle \int_{m-0}^{M} \left(\int_m^M \left(f_2 \circ g^{-1}\right)^{(k)} (g(t)) (g(\lambda) - g(t))^k dt \right) d\langle E_\lambda x, x\rangle \right] \right\} \right\} \Bigg|$$

$$\leq \frac{\|g\|_{\infty,[m,M]}^{n-1} \|g'\|_{\infty,[m,M]}}{(n+1)!(M-m)} \left[\|f_2(A)\| \left\| \left(f_1 \circ g^{-1}\right)^{(n)} \circ g \right\|_{\infty,[m,M]} + \right.$$

$$\left. \|f_1(A)\| \left\| \left(f_2 \circ g^{-1}\right)^{(n)} \circ g \right\|_{\infty,[m,M]} \right] \left[\left\|(M1_H - A)^{n+1}\right\| + \left\|(A - m1_H)^{n+1}\right\| \right]. \tag{10.14}$$

Proof Call $l_i = f_i \circ g^{-1}$, $i = 1, 2$. Then $l_i, l_i', \ldots, l_i^{(n)}$ are continuous from $g([a,b])$ into $f_i([a,b])$, $i = 1, 2$. Hence $\left(f_i \circ g^{-1}\right)^{(n)} \circ g \in C([a,b])$, $i = 1, 2$. Here $\{E_\lambda\}_\lambda$ is the spectral family of A.

Next we use Theorem 26.9 of [1], p. 404. We have that ($i = 1, 2$)

$$f_i(\lambda) = \frac{1}{M-m}\int_m^M f_i(t)\,dt +$$

$$\frac{1}{(M-m)}\left\{\sum_{k=1}^{n-1}\frac{1}{k!}\int_m^M \left(f_i \circ g^{-1}\right)^{(k)}(g(t))\,(g(\lambda)-g(t))^k\,dt\right\} +$$

$$+\frac{1}{(n-1)!\,(M-m)}\int_m^M (g(\lambda)-g(t))^{n-1}\left(f_i \circ g^{-1}\right)^{(n)}(g(t))\,g'(t)\,K(t,\lambda)\,dt, \tag{10.15}$$

$\forall\, \lambda \in [m, M]$,

where

$$K(t,\lambda) := \begin{cases} t-m, & m \le t \le \lambda \le M, \\ t-M, & m \le \lambda < t \le M. \end{cases} \tag{10.16}$$

By applying the spectral representation theorem on (10.15), i.e. integrating against E_λ over $[m, M]$, see (10.4), we obtain:

$$f_i(A) = \left(\frac{1}{M-m}\int_m^M f_i(t)\,dt\right)1_H +$$

$$\frac{1}{(M-m)}\left\{\sum_{k=1}^{n-1}\frac{1}{k!}\int_{m-0}^M\left(\int_m^M \left(f_i \circ g^{-1}\right)^{(k)}(g(t))\,(g(\lambda)-g(t))^k\,dt\right)dE_\lambda\right\}$$

$$+\frac{1}{(n-1)!\,(M-m)}\cdot$$

$$\int_{m-0}^M\left(\int_m^M (g(\lambda)-g(t))^{n-1}\left(f_i \circ g^{-1}\right)^{(n)}(g(t))\,g'(t)\,K(t,\lambda)\,dt\right)dE_\lambda, \tag{10.17}$$

$i = 1, 2$.

We notice that

$$f_1(A)\,f_2(A) = f_2(A)\,f_1(A), \tag{10.18}$$

to be used next.

Hence it holds

$$f_2(A)\,f_1(A) = \left(\frac{1}{M-m}\int_m^M f_1(t)\,dt\right)f_2(A) + \tag{10.19}$$

$$\frac{1}{(M-m)}\left\{\sum_{k=1}^{n-1}\frac{1}{k!}f_2(A)\int_{m-0}^{M}\left(\int_m^M\left(f_1\circ g^{-1}\right)^{(k)}(g(t))(g(\lambda)-g(t))^k dt\right)dE_\lambda\right\}$$

$$+\frac{1}{(n-1)!(M-m)}f_2(A)\cdot$$

$$\int_{m-0}^{M}\left(\int_m^M (g(\lambda)-g(t))^{n-1}\left(f_1\circ g^{-1}\right)^{(n)}(g(t))\,g'(t)\,K(t,\lambda)\,dt\right)dE_\lambda,$$

and

$$f_1(A)\,f_2(A)=\left(\frac{1}{M-m}\int_m^M f_2(t)\,dt\right)f_1(A)+$$

$$\frac{1}{(M-m)}\left\{\sum_{k=1}^{n-1}\frac{1}{k!}f_1(A)\int_{m-0}^{M}\left(\int_m^M\left(f_2\circ g^{-1}\right)^{(k)}(g(t))(g(\lambda)-g(t))^k dt\right)dE_\lambda\right\}$$

$$+\frac{1}{(n-1)!(M-m)}f_1(A)\cdot$$

$$\int_{m-0}^{M}\left(\int_m^M (g(\lambda)-g(t))^{n-1}\left(f_2\circ g^{-1}\right)^{(n)}(g(t))\,g'(t)\,K(t,\lambda)\,dt\right)dE_\lambda. \tag{10.20}$$

Here from now on we consider $x\in H:\|x\|=1$; immediately we get

$$\int_{m-0}^{M} d\langle E_\lambda x,x\rangle=1.$$

Then it holds $(i=1,2)$

$$\langle f_i(A)\,x,x\rangle=\left(\frac{1}{M-m}\int_m^M f_i(t)\,dt\right)+ \tag{10.21}$$

$$\frac{1}{(M-m)}\left\{\sum_{k=1}^{n-1}\frac{1}{k!}\int_{m-0}^{M}\left(\int_m^M\left(f_i\circ g^{-1}\right)^{(k)}(g(t))(g(\lambda)-g(t))^k dt\right)d\langle E_\lambda x,x\rangle\right\}$$

$$+\frac{1}{(n-1)!(M-m)}\cdot$$

$$\int_{m-0}^{M}\left(\int_m^M (g(\lambda)-g(t))^{n-1}\left(f_i\circ g^{-1}\right)^{(n)}(g(t))\,g'(t)\,K(t,\lambda)\,dt\right)d\langle E_\lambda x,x\rangle.$$

It follows that

$$\langle f_2(A)x,x\rangle\langle f_1(A)x,x\rangle=\left(\frac{1}{M-m}\int_m^M f_1(t)\,dt\right)\langle f_2(A)x,x\rangle+\frac{1}{(M-m)}\cdot$$

$$\left\{\sum_{k=1}^{n-1}\frac{1}{k!}\langle f_2(A)x,x\rangle\int_{m-0}^M\left(\int_m^M\left(f_1\circ g^{-1}\right)^{(k)}(g(t))(g(\lambda)-g(t))^k\,dt\right)d\langle E_\lambda x,x\rangle\right\}$$

$$+\frac{1}{(n-1)!(M-m)}\langle f_2(A)x,x\rangle\cdot \tag{10.22}$$

$$\int_{m-0}^M\left(\int_m^M(g(\lambda)-g(t))^{n-1}\left(f_1\circ g^{-1}\right)^{(n)}(g(t))\,g'(t)\,K(t,\lambda)\,dt\right)d\langle E_\lambda x,x\rangle,$$

and

$$\langle f_1(A)x,x\rangle\langle f_2(A)x,x\rangle=\left(\frac{1}{M-m}\int_m^M f_2(t)\,dt\right)\langle f_1(A)x,x\rangle+\frac{1}{(M-m)}\cdot$$

$$\left\{\sum_{k=1}^{n-1}\frac{1}{k!}\langle f_1(A)x,x\rangle\int_{m-0}^M\left(\int_m^M\left(f_2\circ g^{-1}\right)^{(k)}(g(t))(g(\lambda)-g(t))^k\,dt\right)d\langle E_\lambda x,x\rangle\right\}$$

$$+\frac{1}{(n-1)!(M-m)}\langle f_1(A)x,x\rangle\cdot \tag{10.23}$$

$$\int_{m-0}^M\left(\int_m^M(g(\lambda)-g(t))^{n-1}\left(f_2\circ g^{-1}\right)^{(n)}(g(t))\,g'(t)\,K(t,\lambda)\,dt\right)d\langle E_\lambda x,x\rangle.$$

Furthermore we obtain

$$\langle f_1(A)f_2(A)x,x\rangle=\left(\frac{1}{M-m}\int_m^M f_1(t)\,dt\right)\langle f_2(A)x,x\rangle+\frac{1}{(M-m)}\cdot$$

$$\left\{\sum_{k=1}^{n-1}\frac{1}{k!}\left\langle\left(f_2(A)\int_{m-0}^M\left(\int_m^M\left(f_1\circ g^{-1}\right)^{(k)}(g(t))(g(\lambda)-g(t))^k\,dt\right)dE_\lambda\right)x,x\right\rangle\right\}$$

$$+\frac{1}{(n-1)!(M-m)}\cdot \tag{10.24}$$

$$\left\langle\left(f_2(A)\int_{m-0}^M\left(\int_m^M(g(\lambda)-g(t))^{n-1}\left(f_1\circ g^{-1}\right)^{(n)}(g(t))\,g'(t)\,K(t,\lambda)\,dt\right)dE_\lambda\right)x,x\right\rangle,$$

and

$$\langle f_1(A)f_2(A)x,x\rangle=\left(\frac{1}{M-m}\int_m^M f_2(t)\,dt\right)\langle f_1(A)x,x\rangle+\frac{1}{(M-m)}\cdot$$

$$\left\{\sum_{k=1}^{n-1}\frac{1}{k!}\left\langle\left(f_1(A)\int_{m-0}^M\left(\int_m^M\left(f_2\circ g^{-1}\right)^{(k)}(g(t))(g(\lambda)-g(t))^k\,dt\right)dE_\lambda\right)x,x\right\rangle\right\}$$

$$+\frac{1}{(n-1)!(M-m)}\cdot \tag{10.25}$$

$$\left\langle \left(f_1(A) \int_{m-0}^{M} \left(\int_m^M (g(\lambda) - g(t))^{n-1} \left(f_2 \circ g^{-1} \right)^{(n)} (g(t))\, g'(t)\, K(t,\lambda)\, dt \right) dE_\lambda \right) x, x \right\rangle.$$

By (10.22)–(10.24) we obtain

$$E := \langle f_1(A) f_2(A) x, x\rangle - \langle f_1(A) x, x\rangle \langle f_2(A) x, x\rangle = \frac{1}{(M-m)}\cdot$$

$$\left\{ \sum_{k=1}^{n-1} \frac{1}{k!} \left[\left\langle \left(f_2(A) \int_{m-0}^{M} \left(\int_m^M \left(f_1 \circ g^{-1} \right)^{(k)} (g(t)) (g(\lambda) - g(t))^k dt \right) dE_\lambda \right) x, x \right\rangle \right.\right.$$

$$\left.\left. - \langle f_2(A) x, x\rangle \int_{m-0}^{M} \left(\int_m^M \left(f_1 \circ g^{-1} \right)^{(k)} (g(t)) (g(\lambda) - g(t))^k dt \right) d\langle E_\lambda x, x\rangle \right] \right\}$$

$$+ \frac{1}{(n-1)!\,(M-m)}\cdot$$

$$\left[\left\langle \left(f_2(A) \int_{m-0}^{M} \left(\int_m^M (g(\lambda) - g(t))^{n-1} \left(f_1 \circ g^{-1} \right)^{(n)} (g(t))\, g'(t)\, K(t,\lambda)\, dt \right) dE_\lambda \right) x, x \right\rangle \right. \tag{10.26}$$

$$- \langle f_2(A) x, x\rangle \cdot$$

$$\left. \int_{m-0}^{M} \left(\int_m^M (g(\lambda) - g(t))^{n-1} \left(f_1 \circ g^{-1} \right)^{(n)} (g(t))\, g'(t)\, K(t,\lambda)\, dt \right) d\langle E_\lambda x, x\rangle \right],$$

and by (10.23)–(10.25) we have

$$E = \frac{1}{(M-m)}\cdot$$

$$\left\{ \sum_{k=1}^{n-1} \frac{1}{k!} \left[\left\langle \left(f_1(A) \int_{m-0}^{M} \left(\int_m^M \left(f_2 \circ g^{-1} \right)^{(k)} (g(t)) (g(\lambda) - g(t))^k dt \right) dE_\lambda \right) x, x \right\rangle \right.\right.$$

$$\left.\left. - \langle f_1(A) x, x\rangle \int_{m-0}^{M} \left(\int_m^M \left(f_2 \circ g^{-1} \right)^{(k)} (g(t)) (g(\lambda) - g(t))^k dt \right) d\langle E_\lambda x, x\rangle \right] \right\}$$

$$+ \frac{1}{(n-1)!\,(M-m)}\cdot$$

$$\left[\left\langle \left(f_1(A) \int_{m-0}^{M} \left(\int_m^M (g(\lambda) - g(t))^{n-1} \left(f_2 \circ g^{-1} \right)^{(n)} (g(t))\, g'(t)\, K(t,\lambda)\, dt \right) dE_\lambda \right) x, x \right\rangle \right.$$

$$- \langle f_1(A) x, x\rangle \cdot$$

$$\left. \int_{m-0}^{M} \left(\int_m^M (g(\lambda) - g(t))^{n-1} \left(f_2 \circ g^{-1} \right)^{(n)} (g(t))\, g'(t)\, K(t,\lambda)\, dt \right) d\langle E_\lambda x, x\rangle \right]. \tag{10.27}$$

Consequently, by adding (10.26) and (10.27), we get that

$$2E = \frac{1}{(M-m)} \cdot \tag{10.28}$$

$$\left\{ \sum_{k=1}^{n-1} \frac{1}{k!} \left\{ \left[\left\langle \left(f_2(A) \int_{m-0}^{M} \left(\int_{m}^{M} \left(f_1 \circ g^{-1} \right)^{(k)} (g(t)) (g(\lambda) - g(t))^k dt \right) dE_\lambda \right) x, x \right\rangle \right.\right.\right.$$

$$\left. - \langle f_2(A) x, x \rangle \int_{m-0}^{M} \left(\int_{m}^{M} \left(f_1 \circ g^{-1} \right)^{(k)} (g(t)) (g(\lambda) - g(t))^k dt \right) d \langle E_\lambda x, x \rangle \right] +$$

$$\left[\left\langle \left(f_1(A) \int_{m-0}^{M} \left(\int_{m}^{M} \left(f_2 \circ g^{-1} \right)^{(k)} (g(t)) (g(\lambda) - g(t))^k dt \right) dE_\lambda \right) x, x \right\rangle \right.$$

$$\left.\left.\left. - \langle f_1(A) x, x \rangle \int_{m-0}^{M} \left(\int_{m}^{M} \left(f_2 \circ g^{-1} \right)^{(k)} (g(t)) (g(\lambda) - g(t))^k dt \right) d \langle E_\lambda x, x \rangle \right] \right\} \right\}$$

$$+ \frac{1}{(n-1)!(M-m)} \cdot$$

$$\left\{ \left[\left\langle \left(f_2(A) \int_{m-0}^{M} \left(\int_{m}^{M} (g(\lambda) - g(t))^{n-1} \left(f_1 \circ g^{-1} \right)^{(n)} (g(t)) g'(t) K(t, \lambda) dt \right) dE_\lambda \right) x, x \right\rangle \right.\right.$$

$$- \langle f_2(A) x, x \rangle \cdot$$

$$\left. \int_{m-0}^{M} \left(\int_{m}^{M} (g(\lambda) - g(t))^{n-1} \left(f_1 \circ g^{-1} \right)^{(n)} (g(t)) g'(t) K(t, \lambda) dt \right) d \langle E_\lambda x, x \rangle \right] +$$

$$\left[\left\langle \left(f_1(A) \int_{m-0}^{M} \left(\int_{m}^{M} (g(\lambda) - g(t))^{n-1} \left(f_2 \circ g^{-1} \right)^{(n)} (g(t)) g'(t) K(t, \lambda) dt \right) dE_\lambda \right) x, x \right\rangle \right.$$

$$- \langle f_1(A) x, x \rangle \cdot$$

$$\left.\left. \int_{m-0}^{M} \left(\int_{m}^{M} (g(\lambda) - g(t))^{n-1} \left(f_2 \circ g^{-1} \right)^{(n)} (g(t)) g'(t) K(t, \lambda) dt \right) d \langle E_\lambda x, x \rangle \right] \right\}.$$

We find that

$$\langle f_1(A) f_2(A) x, x \rangle - \langle f_1(A) x, x \rangle \langle f_2(A) x, x \rangle - \frac{1}{2(M-m)} \cdot$$

$$\left\{ \sum_{k=1}^{n-1} \frac{1}{k!} \left\{ \left[\left\langle \left(f_2(A) \int_{m-0}^{M} \left(\int_{m}^{M} \left(f_1 \circ g^{-1} \right)^{(k)} (g(t)) (g(\lambda) - g(t))^k dt \right) dE_\lambda \right) x, x \right\rangle \right.\right.\right.$$

$$\left. - \langle f_2(A) x, x \rangle \int_{m-0}^{M} \left(\int_{m}^{M} \left(f_1 \circ g^{-1} \right)^{(k)} (g(t)) (g(\lambda) - g(t))^k dt \right) d \langle E_\lambda x, x \rangle \right] +$$

$$\left[\left\langle \left(f_1(A) \int_{m-0}^{M} \left(\int_{m}^{M} \left(f_2 \circ g^{-1} \right)^{(k)} (g(t)) (g(\lambda) - g(t))^k dt \right) dE_\lambda \right) x, x \right\rangle \right.$$

$$\left.\left.\left. - \langle f_1(A) x, x \rangle \int_{m-0}^{M} \left(\int_{m}^{M} \left(f_2 \circ g^{-1} \right)^{(k)} (g(t)) (g(\lambda) - g(t))^k dt \right) d \langle E_\lambda x, x \rangle \right] \right\} \right\}$$

$$= \frac{1}{2(n-1)!(M-m)} \cdot$$

$$\left\{ \left[\left\langle \left(f_2(A) \int_{m-0}^{M} \left(\int_{m}^{M} (g(\lambda) - g(t))^{n-1} \left(f_1 \circ g^{-1} \right)^{(n)} (g(t)) g'(t) K(t, \lambda) dt \right) dE_\lambda \right) x, x \right\rangle \right.\right.$$

$$- \langle f_2(A) x, x \rangle \cdot$$

$$\int_{m-0}^{M}\left(\int_{m}^{M}(g(\lambda)-g(t))^{n-1}\left(f_1\circ g^{-1}\right)^{(n)}(g(t))\,g'(t)\,K(t,\lambda)\,dt\right)d\langle E_\lambda x,x\rangle\Bigg]+$$
$$\Bigg[\left\langle\left(f_1(A)\int_{m-0}^{M}\left(\int_{m}^{M}(g(\lambda)-g(t))^{n-1}\left(f_2\circ g^{-1}\right)^{(n)}(g(t))\,g'(t)\,K(t,\lambda)\,dt\right)dE_\lambda\right)x,x\right\rangle$$
$$-\langle f_1(A)x,x\rangle\cdot$$
$$\int_{m-0}^{M}\left(\int_{m}^{M}(g(\lambda)-g(t))^{n-1}\left(f_2\circ g^{-1}\right)^{(n)}(g(t))\,g'(t)\,K(t,\lambda)\,dt\right)d\langle E_\lambda x,x\rangle\Bigg]\Bigg\}=:R. \tag{10.29}$$

Hence we have

$$|R|\leq\frac{1}{2(n-1)!(M-m)}\cdot$$
$$\Bigg\{\Bigg[\left|\left\langle\left(f_2(A)\int_{m-0}^{M}\left(\int_{m}^{M}(g(\lambda)-g(t))^{n-1}\left(f_1\circ g^{-1}\right)^{(n)}(g(t))\,g'(t)\,K(t,\lambda)\,dt\right)dE_\lambda\right)x,x\right\rangle\right|$$
$$+|\langle f_2(A)x,x\rangle|$$
$$\left|\int_{m-0}^{M}\left(\int_{m}^{M}(g(\lambda)-g(t))^{n-1}\left(f_1\circ g^{-1}\right)^{(n)}(g(t))\,g'(t)\,K(t,\lambda)\,dt\right)d\langle E_\lambda x,x\rangle\right|\Bigg]+$$
$$\Bigg[\left|\left\langle\left(f_1(A)\int_{m-0}^{M}\left(\int_{m}^{M}(g(\lambda)-g(t))^{n-1}\left(f_2\circ g^{-1}\right)^{(n)}(g(t))\,g'(t)\,K(t,\lambda)\,dt\right)dE_\lambda\right)x,x\right\rangle\right|$$
$$+|\langle f_1(A)x,x\rangle|\cdot$$
$$\left|\int_{m-0}^{M}\left(\int_{m}^{M}(g(\lambda)-g(t))^{n-1}\left(f_2\circ g^{-1}\right)^{(n)}(g(t))\,g'(t)\,K(t,\lambda)\,dt\right)d\langle E_\lambda x,x\rangle\right|\Bigg]\Bigg\} \tag{10.30}$$

(here notice that

$$\left|\int_{m}^{M}(g(\lambda)-g(t))^{n-1}\left(f_1\circ g^{-1}\right)^{(n)}(g(t))\,g'(t)\,K(t,\lambda)\,dt\right|\leq$$
$$\int_{m}^{M}|g(\lambda)-g(t)|^{n-1}\left|\left(f_1\circ g^{-1}\right)^{(n)}(g(t))\right|\left|g'(t)\right|\,|K(t,\lambda)|\,dt\leq$$
$$\left(\int_{m}^{M}|\lambda-t|^{n-1}\,|K(t,\lambda)|\,dt\right)\|g\|_\infty^{n-1}\left\|\left(f_1\circ g^{-1}\right)^{(n)}\circ g\right\|_\infty\left\|g'\right\|_\infty= \tag{10.31}$$
$$\frac{\|g\|_\infty^{n-1}\left\|g'\right\|_\infty\left\|\left(f_1\circ g^{-1}\right)^{(n)}\circ g\right\|_\infty}{n(n+1)}\left[(M-\lambda)^{n+1}+(\lambda-m)^{n+1}\right])$$
$$\leq\frac{1}{2(n-1)!(M-m)}\cdot$$
$$\Bigg\{\Bigg[\left\|f_2(A)\int_{m-0}^{M}\left(\int_{m}^{M}(g(\lambda)-g(t))^{n-1}\left(f_1\circ g^{-1}\right)^{(n)}(g(t))\,g'(t)\,K(t,\lambda)\,dt\right)dE_\lambda\right\|$$
$$+\left\|f_1(A)\int_{m-0}^{M}\left(\int_{m}^{M}(g(\lambda)-g(t))^{n-1}\left(f_2\circ g^{-1}\right)^{(n)}(g(t))\,g'(t)\,K(t,\lambda)\,dt\right)dE_\lambda\right\|\Bigg]$$
$$+\Bigg[\|f_2(A)\|\frac{\|g\|_\infty^{n-1}\left\|g'\right\|_\infty\left\|\left(f_1\circ g^{-1}\right)^{(n)}\circ g\right\|_\infty}{n(n+1)}\cdot$$

$$\left[\left\langle (M1_H - A)^{n+1} x, x\right\rangle + \left\langle (A - m1_H)^{n+1} x, x\right\rangle\right] +$$

$$\|f_1(A)\| \frac{\|g\|_\infty^{n-1} \|g'\|_\infty \left\|(f_2 \circ g^{-1})^{(n)} \circ g\right\|_\infty}{n(n+1)} \cdot$$

$$\left[\left\langle (M1_H - A)^{n+1} x, x\right\rangle + \left\langle (A - m1_H)^{n+1} x, x\right\rangle\right]\Big\} \tag{10.32}$$

$$\leq \frac{1}{2(n-1)!(M-m)} \cdot$$

$$\Big\{\|f_2(A)\| \left\| \int_{m-0}^{M} \left(\int_m^M (g(\lambda) - g(t))^{n-1} \left(f_1 \circ g^{-1}\right)^{(n)} (g(t)) g'(t) K(t,\lambda) dt \right) dE_\lambda \right\| +$$

$$\|f_1(A)\| \left\| \int_{m-0}^{M} \left(\int_m^M (g(\lambda) - g(t))^{n-1} \left(f_2 \circ g^{-1}\right)^{(n)} (g(t)) g'(t) K(t,\lambda) dt \right) dE_\lambda \right\| \Big]$$

$$+ \left[\frac{\|f_2(A)\| \|g\|_\infty^{n-1} \|g'\|_\infty \left\|(f_1 \circ g^{-1})^{(n)} \circ g\right\|_\infty}{n(n+1)} + \right.$$

$$\left. \frac{\|f_1(A)\| \|g\|_\infty^{n-1} \|g'\|_\infty \left\|(f_2 \circ g^{-1})^{(n)} \circ g\right\|_\infty}{n(n+1)} \right] \cdot$$

$$\left[\left\langle (M1_H - A)^{n+1} x, x\right\rangle + \left\langle (A - m1_H)^{n+1} x, x\right\rangle\right]\Big\} =: (\xi). \tag{10.33}$$

Notice here that

$$\left\| \int_{m-0}^{M} \left(\int_m^M (g(\lambda) - g(t))^{n-1} \left(f_1 \circ g^{-1}\right)^{(n)} (g(t)) g'(t) K(t,\lambda) dt \right) dE_\lambda \right\| =$$

$$\sup_{\|x\|=1} \left| \int_{m-0}^{M} \left(\int_m^M (g(\lambda) - g(t))^{n-1} \left(f_1 \circ g^{-1}\right)^{(n)} (g(t)) g'(t) K(t,\lambda) dt \right) d\langle E_\lambda x, x\rangle \right|$$

$$\leq \frac{\|g\|_\infty^{n-1} \|g'\|_\infty \left\|(f_1 \circ g^{-1})^{(n)} \circ g\right\|_\infty}{n(n+1)} \cdot$$

$$\left[\left\|(M1_H - A)^{n+1}\right\| + \left\|(A - m1_H)^{n+1}\right\|\right]. \tag{10.34}$$

A similar estimate to (10.34) holds for f_2.

Hence we obtain by (10.33), (10.34) that

$$(\xi) \leq \frac{1}{(n-1)!(M-m)} \left[\|f_2(A)\| \frac{\|g\|_\infty^{n-1} \|g'\|_\infty \left\|(f_1 \circ g^{-1})^{(n)} \circ g\right\|_\infty}{n(n+1)} \right.$$

$$\left. + \|f_1(A)\| \frac{\|g\|_\infty^{n-1} \|g'\|_\infty \left\|(f_2 \circ g^{-1})^{(n)} \circ g\right\|_\infty}{n(n+1)} \right] \cdot$$

$$\left[\left\|(M1_H - A)^{n+1}\right\| + \left\|(A - m1_H)^{n+1}\right\|\right] = \tag{10.35}$$

$$\frac{\|g\|_{\infty}^{n-1}\left\|g'\right\|_{\infty}}{(n+1)!\,(M-m)}\cdot$$

$$\left[\|f_2(A)\|\left\|\left(f_1\circ g^{-1}\right)^{(n)}\circ g\right\|_{\infty}+\|f_1(A)\|\left\|\left(f_2\circ g^{-1}\right)^{(n)}\circ g\right\|_{\infty}\right]\cdot$$

$$\left[\left\|(M1_H-A)^{n+1}\right\|+\left\|(A-m1_H)^{n+1}\right\|\right].$$

We have proved that

$$|R|\leq\frac{\|g\|_{\infty}^{n-1}\left\|g'\right\|_{\infty}}{(n+1)!\,(M-m)}\cdot$$

$$\left[\|f_2(A)\|\left\|\left(f_1\circ g^{-1}\right)^{(n)}\circ g\right\|_{\infty}+\|f_1(A)\|\left\|\left(f_2\circ g^{-1}\right)^{(n)}\circ g\right\|_{\infty}\right]\cdot \tag{10.36}$$

$$\left[\left\|(M1_H-A)^{n+1}\right\|+\left\|(A-m1_H)^{n+1}\right\|\right],$$

that is proving the claim.

Above it is $\|\cdot\|_{\infty}=\|\cdot\|_{\infty,[m,M]}$. ∎

We give

Corollary 10.4 ($n=1$ case of Theorem 10.3) *For every* $x\in H:\|x\|=1$, *we obtain that*

$$|\langle f_1(A)\,f_2(A)\,x,x\rangle-\langle f_1(A)\,x,x\rangle\,\langle f_2(A)\,x,x\rangle|\leq\frac{\left\|g'\right\|_{\infty,[m,M]}}{2\,(M-m)}\cdot$$

$$\left[\|f_2(A)\|\left\|\left(f_1\circ g^{-1}\right)'\circ g\right\|_{\infty,[m,M]}+\|f_1(A)\|\left\|\left(f_2\circ g^{-1}\right)'\circ g\right\|_{\infty,[m,M]}\right]\cdot$$

$$\left[\left\|(M1_H-A)^2\right\|+\left\|(A-m1_H)^2\right\|\right]. \tag{10.37}$$

We present

Theorem 10.5 *Here all as in Theorem 10.3. Let* $p,q>1:\frac{1}{p}+\frac{1}{q}=1$. *Then*

$$\langle(\Delta(f_1,f_2;g))(A)\,x,x\rangle\leq$$

$$\frac{\|g\|_{\infty,[m,M]}^{n-1}\left\|g'\right\|_{\infty,[m,M]}}{(n-1)!\,(M-m)}\left(\frac{\Gamma(p(n-1)+1)\,\Gamma(p+1)}{\Gamma(pn+2)}\right)^{\frac{1}{p}}\cdot$$

$$\left[\|f_2(A)\| \left\| \left(f_1 \circ g^{-1}\right)^{(n)} \circ g \right\|_{q,[m,M]} + \|f_1(A)\| \left\| \left(f_2 \circ g^{-1}\right)^{(n)} \circ g \right\|_{q,[m,M]} \right] \cdot$$

$$\left[\left\| (M1_H - A)^{n+\frac{1}{p}} \right\| + \left\| (A - m1_H)^{n+\frac{1}{p}} \right\| \right], \tag{10.38}$$

where Γ *is the gamma function.*

Proof We observe that

$$\left| \int_m^M (g(\lambda) - g(t))^{n-1} \left(f_1 \circ g^{-1}\right)^{(n)} (g(t)) g'(t) K(t,\lambda) dt \right| \leq$$

$$\int_m^M |g(\lambda) - g(t)|^{n-1} \left| \left(f_1 \circ g^{-1}\right)^{(n)} (g(t)) \right| |g'(t)| |K(t,\lambda)| dt \leq$$

$$\|g\|_{\infty,[m,M]}^{n-1} \|g'\|_{\infty,[m,M]} \int_m^M |\lambda - t|^{n-1} \left| \left(f_1 \circ g^{-1}\right)^{(n)} (g(t)) \right| |K(t,\lambda)| dt = \tag{10.39}$$

$$\|g\|_{\infty,[m,M]}^{n-1} \|g'\|_{\infty,[m,M]} \left[\int_m^\lambda (\lambda - t)^{n-1} (t-m) \left| \left(f_1 \circ g^{-1}\right)^{(n)} (g(t)) \right| dt + \right.$$

$$\left. \int_\lambda^M (M-t)(t-\lambda)^{n-1} \left| \left(f_1 \circ g^{-1}\right)^{(n)} (g(t)) \right| dt \right] \leq$$

$$\|g\|_{\infty,[m,M]}^{n-1} \|g'\|_{\infty,[m,M]} \left[\left(\int_m^\lambda (\lambda - t)^{(p(n-1)+1)-1} (t-m)^{(p+1)-1} dt \right)^{\frac{1}{p}} + \right.$$

$$\left. \left(\int_\lambda^M (M-t)^{(p+1)-1} (t-\lambda)^{(p(n-1)+1)-1} dt \right)^{\frac{1}{p}} \right] \left\| \left(f_1 \circ g^{-1}\right)^{(n)} \circ g \right\|_{q,[m,M]} =$$

$$\|g\|_{\infty,[m,M]}^{n-1} \|g'\|_{\infty,[m,M]} \left\| \left(f_1 \circ g^{-1}\right)^{(n)} \circ g \right\|_{q,[m,M]} \cdot$$

$$\left(\frac{\Gamma(p(n-1)+1)\Gamma(p+1)}{\Gamma(pn+2)} \right)^{\frac{1}{p}} \left[(\lambda - m)^{n+\frac{1}{p}} + (M-\lambda)^{n+\frac{1}{p}} \right], \tag{10.40}$$

$\forall \lambda \in [m, M]$.

So we got so far

$$\left|\int_m^M (g(\lambda) - g(t))^{n-1} \left(f_1 \circ g^{-1}\right)^{(n)} (g(t)) g'(t) K(t, \lambda) dt\right| \le$$

$$\|g\|_{\infty,[m,M]}^{n-1} \|g'\|_{\infty,[m,M]} \left\|\left(f_1 \circ g^{-1}\right)^{(n)} \circ g\right\|_{q,[m,M]} \cdot$$

$$\left(\frac{\Gamma(p(n-1)+1)\Gamma(p+1)}{\Gamma(pn+2)}\right)^{\frac{1}{p}} \left[(M-\lambda)^{n+\frac{1}{p}} + (\lambda - m)^{n+\frac{1}{p}}\right], \quad (10.41)$$

$\forall \lambda \in [m, M]$.

Hence it holds

$$\left|\int_{m-0}^M \left(\int_m^M (g(\lambda) - g(t))^{n-1} \left(f_i \circ g^{-1}\right)^{(n)} (g(t)) g'(t) K(t, \lambda) dt\right) d\langle E_\lambda x, x\rangle\right| \le$$

$$\|g\|_{\infty,[m,M]}^{n-1} \|g'\|_{\infty,[m,M]} \left\|\left(f_i \circ g^{-1}\right)^{(n)} \circ g\right\|_{q,[m,M]} \cdot$$

$$\left(\frac{\Gamma(p(n-1)+1)\Gamma(p+1)}{\Gamma(pn+2)}\right)^{\frac{1}{p}} \left[\left\|(M1_H - A)^{n+\frac{1}{p}}\right\| + \left\|(A - m1_H)^{n+\frac{1}{p}}\right\|\right], \quad (10.42)$$

for $i = 1, 2$.

Thus we derive

$$\left\|\int_{m-0}^M \left(\int_m^M (g(\lambda) - g(t))^{n-1} \left(f_i \circ g^{-1}\right)^{(n)} (g(t)) g'(t) K(t, \lambda) dt\right) dE_\lambda\right\| \le$$

$$\|g\|_{\infty,[m,M]}^{n-1} \|g'\|_{\infty,[m,M]} \left\|\left(f_i \circ g^{-1}\right)^{(n)} \circ g\right\|_{q,[m,M]} \cdot$$

$$\left(\frac{\Gamma(p(n-1)+1)\Gamma(p+1)}{\Gamma(pn+2)}\right)^{\frac{1}{p}} \left[\left\|(M1_H - A)^{n+\frac{1}{p}}\right\| + \left\|(A - m1_H)^{n+\frac{1}{p}}\right\|\right], \quad (10.43)$$

for $i = 1, 2$.

Next we use (10.42) and (10.43).

Acting as in the proof of Theorem 10.3 we find that

$$|R| \overset{(10.30)}{\le} \frac{1}{2(n-1)!(M-m)} \cdot$$

$$\left\{2\left[\|f_2(A)\| \|g\|_{\infty,[m,M]}^{n-1} \|g'\|_{\infty,[m,M]} \left\|\left(f_1 \circ g^{-1}\right)^{(n)} \circ g\right\|_{q,[m,M]} \cdot\right.\right.$$

$$\left(\frac{\Gamma(p(n-1)+1)\Gamma(p+1)}{\Gamma(pn+2)}\right)^{\frac{1}{p}}\left[\left\|(M1_H-A)^{n+\frac{1}{p}}\right\|+\left\|(A-m1_H)^{n+\frac{1}{p}}\right\|\right]+ \tag{10.44}$$

$$2\left[\|f_1(A)\|\,\|g\|^{n-1}_{\infty,[m,M]}\left\|g'\right\|_{\infty,[m,M]}\left\|\left(f_2\circ g^{-1}\right)^{(n)}\circ g\right\|_{q,[m,M]}\cdot\right.$$

$$\left.\left(\frac{\Gamma(p(n-1)+1)\Gamma(p+1)}{\Gamma(pn+2)}\right)^{\frac{1}{p}}\left[\left\|(M1_H-A)^{n+\frac{1}{p}}\right\|+\left\|(A-m1_H)^{n+\frac{1}{p}}\right\|\right]\right]\Bigg\}=$$

$$\frac{1}{(n-1)!(M-m)}\left\{\|g\|^{n-1}_{\infty,[m,M]}\left\|g'\right\|_{\infty,[m,M]}\left(\frac{\Gamma(p(n-1)+1)\Gamma(p+1)}{\Gamma(pn+2)}\right)^{\frac{1}{p}}\cdot\right.$$

$$\left[\|f_2(A)\|\left\|\left(f_1\circ g^{-1}\right)^{(n)}\circ g\right\|_{q,[m,M]}+\|f_1(A)\|\left\|\left(f_2\circ g^{-1}\right)^{(n)}\circ g\right\|_{q,[m,M]}\right]\cdot$$

$$\left.\left[\left\|(M1_H-A)^{n+\frac{1}{p}}\right\|+\left\|(A-m1_H)^{n+\frac{1}{p}}\right\|\right]\right\}, \tag{10.45}$$

proving the claim. ■

We give for $n=1$:

Corollary 10.6 (to Theorem 10.5) *It holds*

$$|\langle f_1(A)\,f_2(A)\,x,x\rangle-\langle f_1(A)\,x,x\rangle\langle f_2(A)\,x,x\rangle|\leq\frac{\left\|g'\right\|_{\infty,[m,M]}}{(M-m)(p+1)^{\frac{1}{p}}}\cdot$$

$$\left[\|f_2(A)\|\left\|\left(f_1\circ g^{-1}\right)'\circ g\right\|_{q,[m,M]}+\|f_1(A)\|\left\|\left(f_2\circ g^{-1}\right)'\circ g\right\|_{q,[m,M]}\right]\cdot$$

$$\left[\left\|(M1_H-A)^{1+\frac{1}{p}}\right\|+\left\|(A-m1_H)^{1+\frac{1}{p}}\right\|\right]. \tag{10.46}$$

We continue with

Theorem 10.7 *All as in Theorem 10.3. Then*

$$\langle(\Delta(f_1,f_2;g))(A)\,x,x\rangle\leq\frac{(M-m)^{n-1}}{(n-1)!}\|g\|^{n-1}_{\infty,[m,M]}\left\|g'\right\|_{\infty,[m,M]}\cdot$$

$$\left[\|f_1(A)\|\left\|\left(f_2\circ g^{-1}\right)^{(n)}\circ g\right\|_{1,[m,M]}+\|f_2(A)\|\left\|\left(f_1\circ g^{-1}\right)^{(n)}\circ g\right\|_{1,[m,M]}\right]. \tag{10.47}$$

Proof We observe that

$$\left| \int_m^M (g(\lambda) - g(t))^{n-1} \left(f_i \circ g^{-1}\right)^{(n)} (g(t)) g'(t) K(t, \lambda) dt \right| \leq$$

$$\int_m^M |g(\lambda) - g(t)|^{n-1} \left|g'(t)\right| |K(t, \lambda)| \left| \left(f_i \circ g^{-1}\right)^{(n)} (g(t)) \right| dt \leq$$

$$\|g\|_{\infty,[m,M]}^{n-1} \left\|g'\right\|_{\infty,[m,M]} (M - m)^n \left(\int_m^M \left| \left(f_i \circ g^{-1}\right)^{(n)} (g(t)) \right| dt \right) =$$

$$\|g\|_{\infty,[m,M]}^{n-1} \left\|g'\right\|_{\infty,[m,M]} (M - m)^n \left\| \left(f_i \circ g^{-1}\right)^{(n)} \circ g \right\|_{1,[m,M]}, \quad i = 1, 2. \quad (10.48)$$

Hence it holds ($i = 1, 2$)

$$\left| \int_{m-0}^M \left(\int_m^M (g(\lambda) - g(t))^{n-1} \left(f_i \circ g^{-1}\right)^{(n)} (g(t)) g'(t) K(t, \lambda) dt \right) d \langle E_\lambda x, x \rangle \right| \leq$$

$$\|g\|_{\infty,[m,M]}^{n-1} \left\|g'\right\|_{\infty,[m,M]} (M - m)^n \left\| \left(f_i \circ g^{-1}\right)^{(n)} \circ g \right\|_{1,[m,M]}, \quad (10.49)$$

the last is valid since

$$\int_{m-0}^M d \langle E_\lambda x, x \rangle = 1, \text{ for } x \in H : \|x\| = 1. \quad (10.50)$$

Therefore it holds

$$\left\| \int_{m-0}^M \left(\int_m^M (g(\lambda) - g(t))^{n-1} \left(f_i \circ g^{-1}\right)^{(n)} (g(t)) g'(t) K(t, \lambda) dt \right) dE_\lambda \right\| \leq$$

$$\|g\|_{\infty,[m,M]}^{n-1} \left\|g'\right\|_{\infty,[m,M]} (M - m)^n \left\| \left(f_i \circ g^{-1}\right)^{(n)} \circ g \right\|_{1,[m,M]}, \quad (10.51)$$

for $i = 1, 2$.

Acting as in the proof of Theorem 10.3 we find that

$$|R| \overset{\text{(by (10.30), (10.49), (10.51))}}{\leq} \frac{1}{2(n-1)!(M-m)} \cdot$$

$$\Bigg\{ 2 \|f_2(A)\| \|g\|_{\infty,[m,M]}^{n-1} \left\|g'\right\|_{\infty,[m,M]} (M - m)^n \left\| \left(f_1 \circ g^{-1}\right)^{(n)} \circ g \right\|_{1,[m,M]} +$$

$$2 \|f_1(A)\| \|g\|_{\infty,[m,M]}^{n-1} \|g'\|_{\infty,[m,M]} (M-m)^n \left\| \left(f_2 \circ g^{-1}\right)^{(n)} \circ g \right\|_{1,[m,M]} \Bigg\} = \tag{10.52}$$

$$\frac{(M-m)^{n-1}}{(n-1)!} \|g\|_{\infty,[m,M]}^{n-1} \|g'\|_{\infty,[m,M]} \cdot$$

$$\left[\|f_2(A)\| \left\| \left(f_1 \circ g^{-1}\right)^{(n)} \circ g \right\|_{1,[m,M]} + \|f_1(A)\| \left\| \left(f_2 \circ g^{-1}\right)^{(n)} \circ g \right\|_{1,[m,M]} \right],$$

proving the claim. ■

We finish this section with

Corollary 10.8 (to Theorem 10.7, $n = 1$) *It holds*

$$|\langle f_1(A) f_2(A) x, x\rangle - \langle f_1(A) x, x\rangle \langle f_2(A) x, x\rangle| \leq \|g'\|_{\infty,[m,M]} \cdot$$

$$\left[\|f_1(A)\| \left\| \left(f_2 \circ g^{-1}\right)' \circ g \right\|_{1,[m,M]} + \|f_2(A)\| \left\| \left(f_1 \circ g^{-1}\right)' \circ g \right\|_{1,[m,M]} \right]. \tag{10.53}$$

10.4 Applications

We give

Theorem 10.9 *Let $f_1, f_2 \in C'([a,b])$ with $[m,M] \subset (a,b)$, $m < M$. Here A is a selfadjoint linear operator on the Hilbert space H with spectrum $Sp(A) \subseteq [m,M]$. We consider any $x \in H : \|x\| = 1$, and $\rho > 0 : M < \ln \rho$.*

Then

$$|\langle f_1(A) f_2(A) x, x\rangle - \langle f_1(A) x, x\rangle \langle f_2(A) x, x\rangle| \leq \frac{e^M}{2(M-m)\rho}$$

$$\left[\|f_2(A)\| \left\| (f_1 \circ \ln \rho t)' \circ \frac{e^t}{\rho} \right\|_{\infty,[m,M]} + \|f_1(A)\| \left\| (f_2 \circ \ln \rho t)' \circ \frac{e^t}{\rho} \right\|_{\infty,[m,M]} \right]$$

$$\left[\left\|(M1_H - A)^2\right\| + \left\|(A - m1_H)^2\right\| \right]. \tag{10.54}$$

Proof Apply Corollary 10.4 for $g(t) = \frac{e^t}{\rho}$. ■

We continue with

Theorem 10.10 *All as in Theorem 10.9. Let* $p, q > 1 : \frac{1}{p} + \frac{1}{q} = 1$. *Then*

$$|\langle f_1(A) f_2(A) x, x\rangle - \langle f_1(A) x, x\rangle \langle f_2(A) x, x\rangle| \leq \frac{e^M}{(M-m)(p+1)^{\frac{1}{p}} \rho}$$

$$\left[\|f_2(A)\| \left\|(f_1 \circ \ln \rho t)' \circ \frac{e^t}{\rho}\right\|_{q,[m,M]} + \|f_1(A)\| \left\|(f_2 \circ \ln \rho t)' \circ \frac{e^t}{\rho}\right\|_{q,[m,M]}\right]$$

$$\left[\left\|(M 1_H - A)^{1+\frac{1}{p}}\right\| + \left\|(A - m 1_H)^{1+\frac{1}{p}}\right\|\right]. \tag{10.55}$$

Proof Use of Corollary 10.6 and $g(t) = \frac{e^t}{\rho}$; $\rho > 0$, $M < \ln \rho$. ∎

We finish chapter with

Theorem 10.11 *Here all as in Theorem 10.9. Then*

$$|\langle f_1(A) f_2(A) x, x\rangle - \langle f_1(A) x, x\rangle \langle f_2(A) x, x\rangle| \leq \frac{e^M}{\rho} \tag{10.56}$$

$$\left[\|f_1(A)\| \left\|(f_2 \circ \ln \rho t)' \circ \frac{e^t}{\rho}\right\|_{1,[m,M]} + \|f_2(A)\| \left\|(f_1 \circ \ln \rho t)' \circ \frac{e^t}{\rho}\right\|_{1,[m,M]}\right].$$

Proof Use of Corollary 10.8. ∎

References

1. G.A. Anastassiou, *Intelligent Mathematics: Computational Analysis* (Springer, New York, 2011)
2. G. Anastassiou, *Most General Self Adjoint Operator Chebyshev-Grüss Inequalities* (2016, submitted)
3. P.L. Čebyšev, *Sur les expressions approximatives des intégrales définies par les autres proses entre les mêmes limites*. Proc. Math. Soc. Charkov **2**, 93–98 (1882)
4. S.S. Dragomir, *Inequalities for Functions of selfadjoint Operators on Hilbert Spaces* (2011), http://ajmaa.org/RGMIA/monographs/InFuncOp.pdf
5. S. Dragomir, *Operator Inequalities of Ostrowski and Trapezoidal Type* (Springer, New York, 2012)
6. T. Furuta, J. Mićić Hot, J. Pečarić, Y. Seo, *Mond-Pečarić Method in Operator Inequalities. Inequalities for Bounded Selfadjoint Operators on a Hilbert Space* (Element, Zagreb, 2005)
7. G. Grüss, Über das Maximum des absoluten Betrages von $\left[\left(\frac{1}{b-a}\right)\int_a^b f(x) g(x) dx - \left(\frac{1}{(b-a)^2}\int_a^b f(x) dx \int_a^b g(x) dx\right)\right]$. Math. Z. **39**, 215–226 (1935)
8. G. Helmberg, *Introduction to Spectral Theory in Hilbert Space* (Wiley, New York, 1969)

Chapter 11
About a Fractional Means Inequality

Here we present an interesting fractional means scalar inequality. It follows [2].

We make

Remark 11.1 Let $\nu > 0$, $n := \lceil \nu \rceil$ ($\lceil \cdot \rceil$ ceiling of the number), $f(\cdot, y) \in AC^n([a, b])$, $\forall\, y \in [c, d]$ (it means $\frac{\partial^{n-1} f(\cdot, y)}{\partial x^{n-1}} \in AC([a, b])$, $\forall\, y \in [c, d]$). Then the left Caputo partial fractional derivative with respect to x, is given by (see [1], p. 270)

$$\frac{\partial_{*a}^{\nu} f(x, y)}{\partial x^{\nu}} = \frac{1}{\Gamma(n-\nu)} \int_a^x (x-t)^{n-\nu-1} \frac{\partial^n f(t, y)}{\partial x^n} dt, \tag{11.1}$$

$\forall\, y \in [c, d]$, and it exists almost everywhere for x in $[a, b]$, Γ denotes the gamma function.

Then, we get the left Caputo fractional Taylor formula ([3], p. 54)

$$f(x, y) = \sum_{k=0}^{n-1} \frac{\partial^k f(a, y)}{\partial x^k} (x-a)^k + \frac{1}{\Gamma(\nu)} \int_a^x (x-t)^{\nu-1} \frac{\partial_{*a}^{\nu} f(t, y)}{\partial x^{\nu}} dt, \tag{11.2}$$

$\forall\, x \in [a, b]$, for each $y \in [c, d]$.

Above $\left(\int_a^x (x-t)^{\nu-1} \frac{\partial_{*a}^{\nu} f(t, y)}{\partial x^{\nu}} dt\right) \in AC^n([a, b])$, $\forall\, y \in [c, d]$.

Let now $f(x, \cdot) \in AC^n([c, d])$, $\forall\, x \in [a, b]$ (it means $\frac{\partial^{n-1} f(x, \cdot)}{\partial y^{n-1}} \in AC([c, d])$, $\forall\, x \in [a, b]$). Then the left Caputo partial fractional derivative with respect to y, is given by

$$\frac{\partial_{*c}^{\nu} f(x, y)}{\partial y^{\nu}} = \frac{1}{\Gamma(n-\nu)} \int_c^y (y-s)^{n-\nu-1} \frac{\partial^n f(x, s)}{\partial y^n} ds, \tag{11.3}$$

$\forall\, x \in [a, b]$, and it exists almost everywhere for y in $[c, d]$.

G.A. Anastassiou, *Intelligent Comparisons II: Operator Inequalities and Approximations*, Studies in Computational Intelligence 699,
DOI 10.1007/978-3-319-51475-8_11

Then, we get the left Caputo fractional Taylor formula

$$f(x,y)=\sum_{k=0}^{n-1}\frac{\partial^k f(x,c)}{\partial y^k}(y-c)^k+\frac{1}{\Gamma(\nu)}\int_c^y (y-s)^{\nu-1}\frac{\partial_{*c}^{\nu} f(x,s)}{\partial y^{\nu}}ds, \quad (11.4)$$

$\forall\, y\in[c,d]$, for each $x\in[a,b]$.

Above $\left(\int_c^y (y-s)^{\nu-1}\frac{\partial_{*c}^{\nu} f(x,s)}{\partial y^{\nu}}ds\right)\in AC^n([c,d])$, $\forall\, x\in[a,b]$.

Assume

$$\frac{\partial^k f(a,y)}{\partial x^k}=0, \text{ for } k=1,\dots,n-1,\ \ \forall\, y\in[c,d], \quad (11.5)$$

we get

$$f(x,y)-f(a,y)=\frac{1}{\Gamma(\nu)}\int_a^x (x-t)^{\nu-1}\frac{\partial_{*a}^{\nu} f(t,y)}{\partial x^{\nu}}dt. \quad (11.6)$$

Additionally assume $f(a,y)=0$, $\forall\, y\in[c,d]$, then

$$f(x,y)=\frac{1}{\Gamma(\nu)}\int_a^x (x-t)^{\nu-1}\frac{\partial_{*a}^{\nu} f(t,y)}{\partial x^{\nu}}dt, \quad (11.7)$$

$\forall\, y\in[c,d]$, $\forall\, x\in[a,b]$.

Assume

$$\frac{\partial^k f(x,c)}{\partial y^k}=0, \text{ for } k=1,\dots,n-1,\ \ \forall\, x\in[a,b], \quad (11.8)$$

we get

$$f(x,y)-f(x,c)=\frac{1}{\Gamma(\nu)}\int_c^y (y-s)^{\nu-1}\frac{\partial_{*c}^{\nu} f(x,s)}{\partial y^{\nu}}ds, \quad (11.9)$$

$\forall\, y\in[c,d]$, $\forall\, x\in[a,b]$.

Additionally assume that $f(x,c)=0$, $\forall\, x\in[a,b]$, then

$$f(x,y)=\frac{1}{\Gamma(\nu)}\int_c^y (y-s)^{\nu-1}\frac{\partial_{*c}^{\nu} f(x,s)}{\partial y^{\nu}}ds, \quad (11.10)$$

$\forall\, y\in[c,d]$, $\forall\, x\in[a,b]$.

Assuming (11.5) and (11.8), we get

$$2f(x,y)-f(a,y)-f(x,c)=$$

$$\frac{1}{\Gamma(\nu)}\left\{\int_a^x (x-t)^{\nu-1}\frac{\partial_{*a}^{\nu} f(t,y)}{\partial x^{\nu}}dt+\int_c^y (y-s)^{\nu-1}\frac{\partial_{*c}^{\nu} f(x,s)}{\partial y^{\nu}}ds\right\}, \quad (11.11)$$

$\forall\, x\in[a,b]$, $\forall\, y\in[c,d]$.

Additionally assume that $f(a, y) = 0, \forall y \in [c, d]$, and $f(x, c) = 0, \forall x \in [a, b]$, we obtain

$$f(x, y) = \frac{1}{2\Gamma(\nu)} \left\{ \int_a^x (x - t)^{\nu-1} \frac{\partial_{*a}^{\nu} f(t, y)}{\partial x^{\nu}} dt + \int_c^y (y - s)^{\nu-1} \frac{\partial_{*c}^{\nu} f(x, s)}{\partial y^{\nu}} ds \right\}, \quad (11.12)$$

$\forall x \in [a, b], \forall y \in [c, d]$.

We can rewrite (11.11) as follows:

$$f(x, y) - \left(\frac{f(a, y) + f(x, c)}{2} \right) =$$

$$\frac{1}{2\Gamma(\nu)} \left\{ \int_a^x (x - t)^{\nu-1} \frac{\partial_{*a}^{\nu} f(t, y)}{\partial x^{\nu}} dt + \int_c^y (y - s)^{\nu-1} \frac{\partial_{*c}^{\nu} f(x, s)}{\partial y^{\nu}} ds \right\}, \quad (11.13)$$

$\forall x \in [a, b], \forall y \in [c, d]$.

If $0 < \nu < 1$, then $n = 1$, and (11.13) is valid without (11.5) and (11.8), which in this case are void conditions.

Call

$$\Delta f(x, y) := f(x, y) - \left(\frac{f(a, y) + f(x, c)}{2} \right). \quad (11.14)$$

Assume $f \in C([a, b] \times [c, d])$, then

$$\int_a^b \int_c^d \Delta f(x, y)\, dxdy = \int_a^b \int_c^d f(x, y)\, dxdy -$$

$$\left(\frac{(b - a) \int_c^d f(a, y)\, dy + (d - c) \int_a^b f(x, c)\, dx}{2} \right). \quad (11.15)$$

Hence it holds

$$\frac{1}{(b - a)(d - c)} \int_a^b \int_c^d \Delta f(x, y)\, dxdy = \frac{1}{(b - a)(d - c)} \int_a^b \int_c^d f(x, y)\, dxdy -$$

$$\left(\frac{\frac{1}{(d-c)} \int_c^d f(a, y)\, dy + \frac{1}{(b-a)} \int_a^b f(x, c)\, dx}{2} \right). \quad (11.16)$$

Assume now that

$$\frac{\partial_{*a}^{\nu} f(x, y)}{\partial x^{\nu}}, \frac{\partial_{*c}^{\nu} f(x, y)}{\partial y^{\nu}} \in C([a, b] \times [c, d]) \quad (11.17)$$

Clearly, it holds

$$|\Delta f(x,y)| \leq$$

$$\frac{1}{2\Gamma(\nu)}\left\{\left|\int_a^x (x-t)^{\nu-1}\left|\frac{\partial_{*a}^{\nu} f(t,y)}{\partial x^{\nu}}\right| dt + \int_c^y (y-s)^{\nu-1}\left|\frac{\partial_{*c}^{\nu} f(x,s)}{\partial y^{\nu}}\right| ds\right\} \leq$$

$$\frac{1}{2\Gamma(\nu)}\left\{\frac{(x-a)^{\nu}}{\nu}\left\|\frac{\partial_{*a}^{\nu} f}{\partial x^{\nu}}\right\|_{\infty} + \frac{(y-c)^{\nu}}{\nu}\left\|\frac{\partial_{*c}^{\nu} f}{\partial y^{\nu}}\right\|_{\infty}\right\} \leq \tag{11.18}$$

$$\frac{1}{2\Gamma(\nu+1)}\left\{(b-a)^{\nu}\left\|\frac{\partial_{*a}^{\nu} f}{\partial x^{\nu}}\right\|_{\infty} + (d-c)^{\nu}\left\|\frac{\partial_{*c}^{\nu} f}{\partial y^{\nu}}\right\|_{\infty}\right\}.$$

That is

$$|\Delta f(x,y)| \leq \frac{1}{2\Gamma(\nu+1)}\left\{(b-a)^{\nu}\left\|\frac{\partial_{*a}^{\nu} f}{\partial x^{\nu}}\right\|_{\infty} + (d-c)^{\nu}\left\|\frac{\partial_{*c}^{\nu} f}{\partial y^{\nu}}\right\|_{\infty}\right\} =: \lambda. \tag{11.19}$$

Hence

$$\frac{1}{(b-a)(d-c)}\left|\int_a^b \int_c^d \Delta f(x,y)\, dxdy\right| \leq$$

$$\frac{1}{(b-a)(d-c)}\int_a^b \int_c^d |\Delta f(x,y)|\, dxdy \leq \lambda.$$

We have derived:

Theorem 11.2 *Let $\nu > 0$, $n := \lceil \nu \rceil$, $f(\cdot, y) \in AC^n([a,b])$, $\forall\, y \in [c,d]$; and $f(x,\cdot) \in AC^n([c,d])$, $\forall\, x \in [a,b]$. Assume $\frac{\partial^k f(a,y)}{\partial x^k} = 0$, for $k = 1, ..., n-1$, $\forall\, y \in [c,d]$; and $\frac{\partial^k f(x,c)}{\partial y^k} = 0$, for $k = 1, ..., n-1$, $\forall\, x \in [a,b]$. Furthermore, assume $f \in C([a,b] \times [c,d])$ and $\frac{\partial_{*a}^{\nu} f(x,y)}{\partial x^{\nu}}, \frac{\partial_{*c}^{\nu} f(x,y)}{\partial y^{\nu}} \in C([a,b] \times [c,d])$. Then*

$$\left|\frac{1}{(b-a)(d-c)}\int_a^b \int_c^d f(x,y)\, dxdy - \left(\frac{\frac{1}{(b-a)}\int_a^b f(x,c)dx + \frac{1}{(d-c)}\int_c^d f(a,y)dy}{2}\right)\right|$$

$$\leq \frac{1}{2\Gamma(\nu+1)}\left\{(b-a)^{\nu}\left\|\frac{\partial_{*a}^{\nu} f}{\partial x^{\nu}}\right\|_{\infty} + (d-c)^{\nu}\left\|\frac{\partial_{*c}^{\nu} f}{\partial y^{\nu}}\right\|_{\infty}\right\}. \tag{11.20}$$

References

1. G.A. Anastassiou, *Fractional Differentiation Inequalities* (Springer, New York, 2009)
2. G. Anastassiou, A fractional means inequality. J. Comput. Anal. Appl. **23**(3), 576–579 (2017)
3. K. Diethelm, *The Analysis of Fractional Differential Equations* (Springer, New York, 2010)

Printed in the United States
By Bookmasters